Elena Ficara
The Form of Truth

Quellen und Studien zur Philosophie

Herausgegeben von
Jens Halfwassen, Dominik Perler
und Michael Quante

Band 145

Elena Ficara
The Form of Truth

Hegel's Philosophical Logic

DE GRUYTER

Work on this book was supported by a Feodor Lynen Fellowship (2018–2019) provided by the Alexander von Humboldt Foundation.

ISBN 978-3-11-099682-1
e-ISBN (PDF) 978-3-11-070371-9
e-ISBN (EPUB) 978-3-11-070381-8
ISSN 0344-8142

Library of Congress Control Number: 2020943165

Bibliographic Information published by the Deutsche Nationalbibliothek
The Deutsche Nationalbibliothek lists this publication in the Deutsche Nationalbibliografie; detailed bibliographic data are available on the internet at http://dnb.dnb.de.

© 2022 Walter de Gruyter GmbH, Berlin/Boston
This volume is text- and page-identical with the hardback published in 2021.
Print and binding: CPI books GmbH, Leck

www.degruyter.com

Acknowledgments

I wrote about some of the questions that are examined in the book (the meaning of natural logic in Hegel, the meaning of form, Hegel's idea of logic revision, the connection between dialectics and dialetheism, the relationship between *consequentia mirabilis* arguments and dialectical inferences) in papers published between 2013 and 2020 (see the list of papers in the bibliography). I am grateful to the Alexander von Humboldt Foundation for generously supporting my research with a Feodor Lynen Fellowship in 2018 and 2019. Versions of the book chapters were given at various seminars, conferences, workshops and tutorials in several places (Paderborn Universität, Technische Universität Berlin, Freie Universität Berlin, Graduate Center CUNY, Klassik Stiftung Weimar, Technische Universität Dresden, Universität Hannover, Universität Münster, Junge Philosophie Oldenburg, University of Bejing, University of Melbourne, University of Milwaukee, Universal Logic Conferences in Rio de Janeiro, Istanbul and Vichy). I am grateful to the people who provided helpful comments on those and other occasions, among them Günter Abel, Jc Beall, Bob Brandom, Franca d'Agostini, Alessandro de Cesaris, Klaus Düsing, Ruth Hagengruber, Sarah Lebock, Nikolay Milkov, Angelica Nuzzo, Volker Peckhaus, Graham Priest, Rainer Schäfer, Pirmin Stekeler-Weithofer, Klaus Vieweg, Juliette Weyand, Knut Weyand.

I am deeply thankful to to my mother, Franca d'Agostini, for illuminating conversations, detailed comments and critique on previous versions of the book and continuous feedback and encouragement in every phase of its development. I also thank the friends and colleagues who were kind enough to read parts and previous versions of the book chapters, among them Jc Beall, Alessandro de Cesaris, Volker Peckhaus and Graham Priest. I am especially thankful to Volker Peckhaus for his steady support of my projects and for his comments and critique on my work, which always force me to make my ideas clear.

Finally, my thanks go to Christoph Schirmer, Marcus Böhm, Mara Weber and Ute Heller of De Gruyter Verlag for helpful guidance through the submission and publication process, and to Michael Quante and Dominik Perler for their helpful comments on the manuscript and its inclusion in the series "Quellen und Studien zur Philosophie".

Contents

Abbreviated References and Citation —— XI

Introduction —— 1
    The present research and extant works on Hegel's logic —— 1
    Hegel within the history of logic —— 4
    The title —— 7
    The content —— 9

I Logic

1   **Terminological preliminaries: *Das Logische* and *die Logik*, *Verstandeslogik* and *Vernunftlogik* —— 12**
1.1   Logic and natural logic —— 12
1.2   Two logics? —— 15

2   **What does it mean to say that "logic coincides with metaphysics"? —— 19**
2.1   Metaphysics as a part of logic —— 20
2.2   Logic and the objectivity of thought —— 21
2.3   Logic and metaphysics. From nature to theory and back —— 24
2.4   From natural logic to philosophical logic —— 26
2.5   Philosophical logic as conceptual logic —— 27

3   **What kind of logic is Hegel's logic? —— 30**
3.1   Transcendental logic? —— 31
3.2   Hegel's logic and contemporary conceptions of "philosophical logic" —— 33
3.3   Conceptuality and the philosophical approach to logic —— 38
3.4   Concepts and the forms of truth —— 39

Summary —— 41

II Form

4   Hegel on the history of formal logic —— 48
4.1   Aristotle —— 48
4.2   Stoic Logic —— 53
4.3   Leibniz —— 54
4.4   Kant —— 57

5   Hegel on logical forms —— 64
5.1   Five theses on forms —— 64
5.2   Against formal logic? —— 66

6   Is Hegel's logic formal? —— 70
6.1   Formalising Hegel's logic? —— 70
6.2   One hypothesis —— 72
6.3   What are logical forms? —— 73

Summary —— 75

III Truth

7   Truth-bearers —— 84
7.1   Hegel and the sentential nature of truth —— 85
7.2   Hegel's critique of the sentential form —— 86
7.3   "The true is the whole"/"The true is the process" —— 87
7.4   *Satz* and *Urteil* —— 91
7.5   *Richtigkeit* and *Wahrheit* —— 93

8   The meaning of "true" —— 97
8.1   Truth as correspondence —— 97
8.2   The relation between rationality and reality —— 101

9   Hegel's concept of truth in contemporary perspective —— 107
9.1   Coherentism or pragmatism? —— 107
9.2   The Aristotelian core of Hegel's theory of truth —— 110
9.3   Hegel and truth in logic —— 113

Summary —— 115

IV Validity

10   **Dialectic from Zeno to Kant —— 123**
10.1   Zeno, Sophists and Heraclitus —— 124
10.2   Plato —— 127
10.3   The Megarians —— 131
10.4   Aristotle —— 134
10.5   The Sceptics —— 137
10.6   Kant —— 141

11   **Hegel's own account of dialectical inferences —— 144**

12   **What is dialectic? —— 148**
12.1   Hegel's account within the history of logic —— 150
12.2   Controversies on the nature of dialectical inferences —— 158
12.3   Truth and Validity —— 161

Summary —— 165

V Contradiction

13   **Conjunction [*Vereinigung*] —— 173**

14   **Negation —— 179**

15   **The Law of Non-Contradiction and the Law of Excluded Middle —— 187**

16   **Hegelian paraconsistentism —— 194**

Summary —— 199

Bibliography —— 202

Index of Names —— 218

Index of Subjects —— 222

Abbreviated References and Citation

Hegel's works are quoted following the *Theorie Werkausgabe*, abbreviated as "Hegel Werke", followed by the indication of the volume and page. If the English translation is available, I also give the page number of the translation. If it is not, the translation is mine. In some cases I have modified the extant English translations, opting for more literal versions, and eliminating possible sources of ambiguities. If the changes are major ones, I explain my motivations in the footnotes. In general, I do not use capital letters for the English equivalent of words such as "Begriff", "Wahrheit", "Idee", words that, in some translations, are written with capital letters ("Notion", "Truth", "Idea"). This habit generates an emphasis that is absent in the German text, and could be misleading.

Kant's works are quoted following the *Königlich Preußische (später Deutsche) Akademie der Wissenschaften* edition (quoted as AA, followed by the indication of volume and page. The *Critique of Pure Reason* is abbreviated, as usual, as A – first edition – and B – second edition, followed by the indication of the page).

Introduction

The present research and extant works on Hegel's logic

This book is a consideration of Hegel's view on logic and basic logical concepts (such as truth, form, validity, contradiction) aiming to assess this view's relevance for contemporary philosophical logic. The literature on Hegel's logic is fairly rich. The subject is dealt with in different perspectives and for different aims, in philosophical or historical approaches, with exegetic or theoretical concern, and using formal or informal methods. The attention to contemporary philosophical logic places the present research closer to those works interested in the link between Hegel's thought and analytical philosophy.[1]

In this context, the first particularity of this book consists in focusing on something that has been generally underrated in the literature: the idea that, for Hegel as well as for Aristotle and many other authors (including Frege), "logic" is the study of *the forms of truth*, i.e. the forms that our thought can (or ought to) assume in searching for truth.[2] As I will explain in this introduction, and elaborate further in the book, the idea of logic as the study of the forms of truth is useful for clarifying some controversial aspects of Hegel's views on logic, and for illustrating the sense in which a reappraisal of Hegel's dialectical logic

---

[1] An account of the literature will be given in the book. In 1976, 75 Gadamer stressed that the works traditionally engaged in reading Hegel from a contemporary point of view normally consider the *Phenomenology of Spirit*, and are interested in examining the meaning of Hegel for epistemology and philosophy of mind, while assessing the relevance of Hegel's logic and metaphysics for contemporary philosophy is traditionally held as thorny. This judgement can now be partially revised. Today there are different works engaged in reading Hegel's logic and metaphysics from an analytical perspective. Among the most recent ones see: Stekeler-Weithofer 1992, Bencivenga 2000, Burbidge 2004, 131–176, Berto 2005, Rockmore 2005, Ameriks/Stolzenberg 2005 (eds.), Redding 2007, Hammer 2007, Nuzzo 2010a (ed.), Butler 2012, Brandom 2014, 1–15, Bordignon 2014, Pippin 2016, Chapter II, Pippin 2019. See also the essays collected in Emundts/Sedgwick 2017, Moyar 2017, Quante/Mooren 2018.

[2] In *Der Gedanke* (English translation Frege 1956, 290) Frege writes: "I assign to logic the task of discovering the laws of truth, not of assertion or thought". In 1897 (now in Frege 1979, 3) Frege defines logic as "the science of the most general laws of truth". He writes "the laws of logic are nothing else than an unfolding of the content of the word 'true' [...] Anyone who has failed to grasp the meaning of this word [...] cannot attain to any clear idea of what the task of logic is". On the continuity between Hegelian, Aristotelian and modern logic see also in more detail Chapters 2 and 3. For clarifying insights on the link between logic and truth see d'Agostini 2011, 115–127 and the unpublished paper "Logic: the alethic account".

may be fruitful for contemporary philosophical debates. More specifically, in this light it becomes possible:
- To understand the precise meaning of Hegel's thoughts on logic and metaphysics, dismissing the dominant view according to which, in Hegel's writings, there would be a confusion between the two disciplines. The concept of truth, commonly intended, is what positively joins logic and metaphysics.[3]
- To highlight the non-psychologistic nature of Hegel's view on logic. The concept of truth is not an exclusively epistemic or mental affair.[4]
- To underline that Hegel's logic, although non-formalised and non-formalistic in principle, is not anti-formal (insofar as it is concerned with the forms of truth).[5]
- To stress that Hegel's view on logic is perfectly in line with the logical tradition and with Frege himself.[6]
- To point out that Hegel's logic, more specifically, corresponds to what today is called *philosophical logic*.[7] As such, it is not incompatible, in principle, with the mathematical appraisal of logic, though including an attention for truth that is not peculiar for mathematical logic.[8]

---

[3] See Part I, Chapter 2. below. Hegel's supposed identification of logic and metaphysics seems to preclude any treatment in terms of modern logic, which is informed by the idea of ontological neutrality. See Peckhaus' reconstruction of the debate in Germany in the second half of the 19th century, when "logic reform meant overcoming the Hegelian identification of logic and metaphysics", Peckhaus 1999, 447. Against the postulate of logic's ontological neutrality see Varzi 2014, 53–80.

[4] Hegel explicitly criticises the subjectivistic and epistemicistic account of logic typical of the logic of his times, openly praising, in this respect, the ancient conception (see Parts I and III below).

[5] See 4.3. and 5.2.

[6] Hegel's critique of traditional logic does not entail a rejection of logic as theory of valid inference. It rather brings Hegel to the discovery of dialectic as the *genuine theory of valid inference* (see Parts II and IV). Even if Hegel criticises the propositional form and emphasises the conceptual nature of dialectical logic, which seems to suggest a major difference between his and the modern view of logic, he also points out, in a way reminiscent of Frege's insights on the same subject, that the "sentence is were truth begins" and that the study of concepts, sentences and inferences should be developed showing their organic unity (see Part III below). On the continuity between Hegelian, Aristotelian and modern logic see also in more detail Chapters 2 and 3, as well as Redding 2007 and 2014, 281–301 and Brandom 2014, 1–15. On the Hegelian assumptions at the basis of Frege's view on logic see Käufer 2005, 259–280. On the influence of Trendelenburg's interpretation of Leibniz on Frege see Gabriel 2008, 115–131.

[7] Even if the expression "philosophical logic" is not always used univocally today, I show in Chapter 3. the continuity between Hegel's use and one specific, but fundamental, meaning of the term, i.e. Russell's idea and definition of "philosophical logic".

[8] See Part I, Chapter 3.

The second peculiarity of the present research is methodological. I tried to reconstruct Hegel's theses on "logic", "form", "truth" etc. not only in his published writings, but with major attention to his *Lectures on the History of Philosophy*, looking at what Hegel writes on ancient and modern authors' conceptions of "logic", "logical form", "dialectical inferences" etc. In so doing, I share a view that is well known among authors belonging to the European tradition of Hegel studies, namely that these *Lectures* are fundamental for understanding the specifically logical meaning of Hegel's thought. They were conceived and held for the first time in Jena in 1805–06, then re-proposed with little to no changes in Heidelberg in 1816–17, before being given regularly in Berlin between 1819 and 1830. They are, as Gadamer highlights in 1976, a true laboratory for Hegel himself, who articulated his conception of dialectic through the interpretation of both classical and modern authors. Moreover, they were evidently addressed to students, and hence their text is exemplarily clear, in contrast to the notoriously obscure published logical writings.

The thesis according to which the *Lectures on the History of Philosophy* are fundamental for understanding the genesis and meaning of Hegel's logical concepts has been explicitly underlined and/or adopted methodologically by several authors in different works.[9] Yet these works do not deal with the significance of Hegel's logic for contemporary philosophical logic. In contrast, the works that do address the question about the relevance of Hegel's logic for contemporary logic do not consider the *Lectures on the History of Philosophy*.[10] In this panorama, my book is motivated by the conviction that Hegel's analysis of the history of philosophy presents fundamental insights for locating Hegel's logic within the history of logic, explaining its link to ancient (Aristotelian and Stoic) and transcendental (Kantian) logic, and so making an assessment of its meaning for contemporary philosophical logic genuinely possible.

Accordingly, I also adhere to a methodological device introduced by Dilthey and adopted by many Hegel scholars,[11] who underline the importance of Hegel's early writings for understanding the motivations and genesis of Hegel's thoughts

---

[9] See Gadamer 1976, Düsing 1976, Id. 1990, 169–191 and Id. 2012, Pöggeler 1990, 42–64, Riedel (ed.) 1990a, Schäfer 2001 among others.

[10] See the essays collected in Marconi (ed.) 1979a, Priest 1989, 388–415, Steckeler-Weithofer 1992, Bencivenga 2000, Berto 2005 among others. Exceptions are d'Agostini 2011b, 121–140, Butler 2012 and Redding 2014, 281–301. However, these authors do not develop a systematic analysis of the logical importance of Hegel's *Lectures on the History of Philosophy*, such as the one I aim at. I deepen these methodological considerations in Part IV.

[11] See Dilthey 1921, Marcuse 1941, Lukács 1973. See also Gadamer 1976, Düsing 1976, Henrich 1965/66, Verra 2007, Berti 2015 and Vieweg 1999 among others.

about logic. Again, these authors' primary task is not a consideration of Hegel's thoughts from the point of view of contemporary philosophical logic. In contrast I stress that the early writings,[12] in particular the Frankfurt's fragments on *Vereinigung*, as well as the Jena's writings: *Differenzschrift* and the *Skeptizismusaufsatz* (writings that are either not translated into English or only partially translated into English) are a crucial reference point for understanding the originality of Hegel's approach with respect to contemporary debates in philosophical logic.

Hegel within the history of logic

In contemporary accounts of the history of logic, Hegel's theses on logic and basic logical notions (such as negation, contradiction, form, truth among others) are the origin of general embarrassment. They are either held to be so incompatible with the tradition of modern formal logic that they are simply not mentioned, or only hinted at as "curious" (Kneale/Kneale 1962, 355), a "mixture of metaphysics and epistemology" (Kneale/Kneale 1962, 355) as involving a "complete rejection of formal logic" and "its substitution with a dialectic which is the product of speculative metaphysics" (Ritter/Gründer/Gabriel (eds.) 1971ff., vol. 5., 358) and thus soon dismissed. A further sign of the embarrassment is, on occasion, the duplication of logical entries in philosophical dictionaries ("logic" and "speculative/dialectical logic"; "negation" and "negation/negativity: Hegel").[13]

Standardly, Hegel's position does not belong to the canon: It is simply not considered, or it is held to belong to a period of logical decadence. For example, according to Bochenski 1978 the period between the 16th and the 19th century is a dead period in the history of logic. Similarly, Boyer distinguishes between three periods in the history of logic: "The history of logic may be divided, with some slight degree of oversimplification, into three stages: (1) Greek logic, (2) Scholastic logic, and (3) mathematical logic" (Boyer 1968, 633). These and similar reconstructions are criticised or revised in more recent accounts. Peckhaus (1999, 434–435) suggests that Boyer's "slight degree of oversimplification [...] enabled him to skip 400 years of logical development and ignore the fact that Kant's transcen-

---

12 See Hegel Werke 1 and 2.
13 See the two entries on "logic" and "speculative/dialectical logic" in Ritter/Gründer/Gabriel (eds.) 1971ff., vol. 5., 357ff. and 389ff. as well as Ritter/Gründer/Gabriel (eds.) 1971ff., vol. 6., 666ff. and 671ff. on "Negation" (in logic) and "Negation/Negativity" (in general in philosophy, and Hegel).

dental logic, Hegel's metaphysics and Mill's inductive logic were called 'logic' too".[14] Thiel 1965, Peckhaus 1997 (as well as Peckhaus 1999), and Gabriel 2008 are explicitly engaged in revising this canon, showing the roots of symbolic logic in the philosophical and logical tradition from Leibniz to Trendelenburg. Expression of a new tendency in the history of logic, also influenced by the arising and development of paraconsistent logics (see Priest and Routley 1984) is the *Handbook of the History of Logic* (Gabbay/Woods (eds.) 2004) in which the authors devote one chapter to Hegel's logic.[15] Also in *The Development of Modern Logic* (Haaparanta 2009), the chapter on the "logical question" (Vikko 2009, 203–221) contains an assessment of Hegel's role in the complex movement of logic reform in the 19th Century known as *die logische Frage* ("the logical question").[16] A complete integration of Hegel into the canon of the history of logic remains, however, a desideratum. More generally, that "the standard presentations of the history of logic ignore the relationship between the philosophical and the mathematical side of its development" is stressed by Peckhaus (1999, 434). I have oriented my research by taking this very relation into account.

Perplexities, omissions and duplications are not completely unjustified. Yet I suggest that, although "logic" in Hegel does denote a wide enterprise, its *core* is perfectly consistent with what we may intend by "logic" nowadays.[17] The methodological line of this book is inspired by the idea that Hegel's contribution to – and his critique of – formal logic can be understood only if we give for granted that, when Hegel spoke of "logic" and mentioned logical concepts, he technically meant something that is not so far from what contemporary logicians mean.

---

14 The *Oxford Handbook of German Philosophy in the Nineteenth Century* devotes one chapter to Nineteenth-Century German logic (Priest 2015, 398–415). In it, Priest does consider Hegel's logic (402ff.), but he also claims that the dialectical cycle "has absolutely nothing to do with inference, and so with the sense of logic in this chapter".
15 It is the chapter written by Burbidge (Burbidge 2004, 131–176).
16 On Hegel and the "logical question" see also Ficara 2015, 39–55.
17 Stekeler-Weithofer 1992, 23 emphasizes that Hegel's logic, in spite of its apparent effusiveness, does correspond to what we mean by "logic" today. On common prejudices against Hegel and his idea of logic see also Stekeler-Weithofer 2016, 3–16. On the breath of topics that Hegel seems to accord to the domain of logic see Siep 2018, 651–798 and Tolley 2019, 73–100. Siep 2018, 790 underlines that Hegel's logic is an all-encompassing systematic inquiry that mediates between different cultural and scientific spheres. For him, such conception of logic is difficult to digest today. However, for Siep it is also beyond doubt that Hegel's logic, so conceived, corresponds to a philosophical account of logic and has, as such, a critical potential against scientific and cultural dogmatisms. For Tolley 2019, 73 f. Hegel's view of logic is characterized by two commitments (which he calls the "over-enrichment" of logic and the "divinization" of its subject matter) that seem to push Hegel away from most contemporary notions of logic.

Accordingly, the general aim of this work is to set the conditions for a genuine integration of Hegel's view on logic and basic logical notions – especially: truth, contradiction, negation, validity, and 'logic' itself – into the canon of contemporary philosophical logic. Such integration is highly fruitful, for many reasons.

First, it is important for exegetical reasons. Many of Hegel's views, if read from the perspective of their postulated "eccentricity" with respect to the tradition of formal logic, do not make any sense, while they gain a clear meaning once the eccentricity assumption is dismissed. Oddly enough, the idea of Hegel's extraneousness to the most canonical development of modern logic plays a certain role in the choice of non-literal translations. For example, a common translation of the passage on Aristotle's logic in the *Lectures on the History of Philosophy* is "Aristotle's *philosophy* is not by any means founded on this relationship of the understanding [the syllogistic forms]" (Hegel Werke 19, 241/Hegel 1892ff., vol. 2, 223). However, in the German text we have "Aristoteles' Logik" and not "Philosophie". It is reasonable to suppose that this shift is due to the conviction, on the part of the interpreter, that when Hegel writes "logic" he does not mean what "logic" is for us, but something else, whereby the more general term "philosophy" is preferable. In the first part of the book I show that Hegel's use of "logic" is perfectly adaptable to our use. More precisely, what he intends by this term corresponds to what we would call *philosophical logic*.

Second, the "normalisation" of Hegel's view on logic is *philosophically* important. Hegel repeatedly stressed that logic is fundamental for theoretical, but also for practical and even political reasons: it is helpful for pursuing what is universally and also personally good, it concerns what intimately interests us. If we accept that what he was talking about was indeed related to logic as we can intend it nowadays, these considerations become more insightful and interesting. They are not simply the result of using "logic" in a loose and vague acceptation, but rather express the positive idea of the usefulness of logic, as formal study of validity, for the shared and personal life of human beings (an idea that is not so far from what Leibniz, or Russell himself, used to think). In this sense, Hegel's work may reveal to be a genuine operation of "logic-empowerment". Yet, if the dominant view is that what Hegel was talking about was not *logic* in the technical meaning of the word, but something else, theses and arguments that, for someone who is interested in logic, are precious, get lost.

Finally, considering Hegel as a genuine interlocutor (perhaps not at the level of Aristotle, or Leibniz, or Frege, but still as an important one) is fruitful for scientific reasons. For example, a consideration of paraconsistentism that ignores Hegel's acquisitions about contradictions would be lacking, in many ways. Generally, paraconsistentists do not ignore Hegel's dialectical approach, even if they do not always take open note of it.

The title

The subtitle of the book refers to the expression "philosophische Logik", coined by Hegel in the *Grundlinien zur Philosophie der Rechts* with reference to the logic of philosophical inquiry, the logic presented by Hegel himself in his logical writings. As I explain in more detail in the first part of the book, contemporary philosophers (Sainsbury 2001, 1 and Lowe 2013, 1) use the same expression in recognition of Bertrand Russell, who, in *Our Knowledge of the External World*, wrote:

> [S]ome kind of knowledge of logical forms, though with most people it is not explicit, is involved in all understanding of discourse. It is the business of philosophical logic to extract this knowledge from its concrete integuments, and to render it explicit and pure. (Russell 2009, 35)

Hegel intends the business of his *philosophische Logik* in very similar terms. For Hegel as well as for Russell, the forms are "facts" deposited in our thought, language and life. Both philosophers also contend that the task of philosophical logic is to "extract the forms" from their "concrete integuments". Hegel writes:

> The forms of thought are [...] displayed and stored in human language [...] [They] permeate every [human] sensation, intuition, desire, need, instinct [...] [They are] the natural element in which human beings [live], indeed [their] own peculiar nature [...] the activity of thought which is at work in all our ideas, purposes, interests and actions is [...] unconsciously busy (natural logic) [...] To focus attention on this logical nature [...] this is the task. (Hegel Werke 5, 20 ff./Hegel 1969, 31 ff.)

In this sense the expression "philosophical logic" is well suited to answer the question "what kind of logic is Hegel's logic?" from a contemporary perspective.

More specifically, as I hinted, the distinctive character of philosophical logic in Hegel's view (as well as in other authors' view) is that logic is strictly, even inextricably, connected to truth. The idea of philosophical logic as a logic of truth is still plausible today.[18] In this sense the expression "the form of truth" is useful to address different aspects of the Hegelian conception of logic.

---

18 The link between logic and truth is addressed in different ways by many classical and contemporary authors. See among others Frege 1979, and Tugendhat 1970. It is the specific subject of d'Agostini 2011, part II.

First, Hegel's logic is informed by the idea according to which a fundamental assumption of the logical work is the realist meaning of truth, and the correspondence of thought with reality. Hegel writes:

> One can appeal to the conception of ordinary logic itself, for it is assumed that, [...] if from given determinations others are inferred, [...] what is inferred is not something external and alien to the object, but rather that it belongs to the object itself, that to the thought there is a correspondent being. (Hegel Werke 5, 45/Hegel 1969, 50 f.)

Here Hegel does not want to say that the inferential forms isolated and considered in logic, for example *modus ponens*, are always and in each case expression of reality (of how things stand), and that an argument such as "If Hillary Clinton is a woman, then she is not reliable. Hillary Clinton is a woman, hence she is not reliable" is expression of how things stand. What Hegel claims is rather that, in individuating the different forms as forms of valid inference, we implicitly assume that they reveal that there is something that corresponds to our thought. According to Hegel, in doing logic we *raise the claim* that the forms we find (extract from the natural logic of language) are forms of truth, the forms that our thought assumes as soon as it is validly engaged in the search for truth.[19]

Second, the expression "the form of truth" refers to a further important aspect of Hegel's idea of logic. Hegel claims that there is a crucial question "no one thinks of investigating", namely "whether these forms [individuated by the logicians] are [in themselves] forms of truth" (Hegel Werke 6, 268/Hegel 1969, 594 f.). The question about the truth of the forms must be asked, for Hegel, and this means that "it is necessary to subject [the forms] to criticism" to determine if they are factually able to express truth, i.e. to give an account of *thought thinking reality*, to genuinely be *forms of thought thinking reality*. So stated, the idea also introduces the possibility (and the need) for a revision of logic, and a criticism of established classical rules.

Finally, the title "The Form of Truth" is to be traced back to what is, for Hegel, the most general form of true thought. For Hegel the forms (and the laws) dealt with in the logic of his times *are not forms and laws of truth*. Contradiction is the only genuine form of truth. Hegel notoriously claims: *contradictio est regula veri* (contradiction is the *norm* of truth), a formulation that fits well with the normative conception of forms Hegel inherits from Kant.[20] In perhaps less well-known passages, Hegel also stresses that contradiction is the *formal character* of truth (Hegel Werke 2, 39). As I show in the last part of the book,

---

[19] This means that, in Hegel, the notion of logic as study of the "forms of thought" is preserved.
[20] I examine the normative meaning of logic in Hegel in the second part of the book.

this conception, which may seem counterintuitive, is a cornerstone of Hegel's dialectical logic. It turns out to be perfectly plausible from the point of view of dialectical and speculative logic as *philosophical logic*. Hegel writes:

> There is a general failure to perceive that, in the case of any knowledge, and any science, what is taken for truth, even as regards content, can only deserve the name of 'truth' when philosophy has had a hand in its production. (Hegel Werke 3, 63/Hegel 1977, 41)

Hegel hints here at philosophical thought as the only condition for the search of (and knowledge of) truth. In stating this, he has in mind an idea of philosophy as sceptical, self-reflexive and critical thought, guided by the ancient sceptical principle that "for every valid argument there is an opposite one that is equally valid".[21] Critically reflecting on (and questioning) every assumption is the condition (the norm) for attaining truth. Hegel derives this view from ancient philosophy – from the Socratic, Platonic and Aristotelian dialectic, as well as from ancient scepticism. In this sense, the expression "the form of truth" refers to *contradiction as norm for attaining truth*. Moreover, the expression implies that *contradiction as form* is the *essential characteristic* of the very concept of truth. If we want to fix the process of our critical, self-reflexive search for (and knowledge of) truth sententially and linguistically, using a finite expression, than this expression will be the contradiction.[22]

The content

The book has five parts, devoted respectively to logic, form, truth, validity and contradiction. At the end of each part and after a consideration of the relation between Hegel's view and contemporary theories, I extensively summarize the contents of each chapter. Here I therefore limit myself to sketching the general themes addressed in the five parts of the book.

The first part (*I: Logic*) is devoted to exploring Hegel's use of the concept of "logic" and his basic views about the nature, reason and aims of logic. I present three Hegelian theses about logic: the distinctions between *das Logische* and *die Logik*, the interplay between intellectual and rational logic, and the connection

---

[21] Sextus Empiricus 1985, 140 and Hegel Werke 2, 230. On scepticism and philosophy in Hegel see, first of all, Vieweg 1999, Id. 2007 and Heidemann 2007. See also Gadamer 1976, Düsing 1976, and the essays collected in Riedel (ed.) 1990a.
[22] In part IV and part V I argue that what Hegel means by "contradiction" can be best grasped formally in terms of $a \leftrightarrow \neg a$.

between logic and metaphysics. The focus is on more or less explicitly shared views about Hegel's logic and its supposed non-governability by means of contemporary logic. Views such as: "Hegel's logic is conceptual and not propositional" or "Hegel's logic is metaphysical and thus has nothing to do with logic commonly intended, which is ontologically neutral", are examined and discussed. The main aim is to discard these views, opening the field for a genuine integration of the Hegelian reflections into the canon of the history of logic and philosophical logic.

The second part (*II: Form*) explores Hegel's view on logical forms and formalisms. Questions such as "is Hegel's logic formal or not?", "is Hegel's view on the formality of logic compatible with both modern and contemporary conceptions?" and finally "is a formal consideration and even a formalisation of Hegel's dialectics completely pointless?" are considered.

In the third part (*III: Truth*) I present Hegel's conception of truth, focusing on two standard questions at the basis of every truth theory in contemporary philosophical logic: "What are the truth-bearers, for Hegel?" and "What does the word "true" mean, for him?". The part closes with a consideration of Hegel's view on truth from the perspective of the link between logic and truth.

The fourth part (*IV: Validity*) is about Hegel's view on "what follows from what" and is motivated by the insight that to assess Hegel's possible contribution to both the history and our actual comprehension of the concept of validity (or logical consequence), it is necessary to understand the exact meaning of dialectical inferences, distinguishing them from other kinds of inferences. In other words, to understand Hegel's notion of validity we have to reflect on Hegel's notion of dialectic.

The last part (*V: Contradiction*) is on the meaning of dialectical contradictions and Hegel's possible contribution to contemporary conceptions of inconsistency. Here I give a closer look at the connectives and correspondent logical laws (Double Negation Elimination, Law of Non Contradiction, Law of Excluded Middle) involved in (or questioned by) contradictions.

I Logic

The forms of thought [are] the natural element in which human beings [live], indeed [their] own peculiar nature [...] To focus attention on this logical nature [...] this is the task. (Hegel Werke 5, 26 f./Hegel 1969, 36 f.)

Some kind of knowledge of logical forms, though with most people it is not explicit, is involved in all understanding of discourse. It is the business of philosophical logic to extract this knowledge from its concrete integuments, and to render it explicit and pure. (Russell 2009, 35)

1 Terminological preliminaries: *Das Logische* and *die Logik*, *Verstandeslogik* and *Vernunftlogik*

1.1 Logic and natural logic

As Hans-Georg Gadamer pointed out (1976, 78), Hegel coins a new expression, which cannot be found before him: "the logical" (*das Logische*). Gadamer suggests that Hegel uses it in the same way that the Greek philosophers used the word *logos*, as an equivalent to "reason", that is: the realm of *concepts* or *forms*, the universal and pure entities constituting and ruling human language and reasoning.[1]

In the Preface to the second edition of the *Science of Logic* Hegel writes:

> The forms of thought are, in the first instance, displayed and stored in human language [...] The Logical [*das Logische*] [is the] natural element [in which] human beings [live], indeed [their] own peculiar nature. (Hegel Werke 5, 26 f., Hegel 1969, 36 f.)

What is interesting here is that logical forms ("the forms of thought") are, for Hegel, objective entities, considerable as "facts" in every respect. These facts occur in a specific field, the field of thought (*das Logische*), which is the distinctive feature (the "natural element") of human beings, and the field in which they live, act and interact.

---

[1] In English translations, the term is often rendered with "logic" (see for instance Hegel 1969, 36 f.), but this could be misleading, as it risks overlooking important philosophical implications. Nuzzo 1997, 41 ff. considers Hegel's distinction between "logic" and "the logical". See also Nuzzo 1992, 193–198 and 281 note 84. Fulda 1965 and Fulda 2006, 25–27 and 32 ff. stresses that "the logical" is the field of Hegel's "first philosophy" or metaphysics. d'Agostini 2000, 95 ff. examines the consequences of Hegel's new use for the relation between logic and metaphysics. Labarrière 1984, 35–41 and more recently Caron 2006, 149–183 propose a theological interpretation of *"das Logische"*. Di Giovanni 2007, 85–87 rejects the theological interpretation, stressing that the expression *"das Logische"*, in Hegel, stands for the field of language and thought that constitutes the subject matter of Hegel's *Science of Logic*. For a consideration of the central role of *"das Logische"* in Hegel's philosophy and in Gadamer's interpretation of Hegel see also Dottori 2006, 423–436 and 530 ff. Abel 1999, 18 ff. distinguishes between two meanings of "logic": logic as theory of valid inference and logic as the forms that are implicitly present in all our actions, thoughts and perceptions. In 1999, 81 ff. Abel also underlines that philosophers such as Plato, Aristotle, Hegel, Frege and Wittgenstein conceived logic also in the second meaning.

1.1 Logic and natural logic — 13

> The activity of thought which is at work in all our ideas, purposes, interests and actions is [...] unconsciously busy (natural logic) [...] To focus attention on this logical nature [...] this is the task. (Hegel Werke 5, 26 f., Hegel 1969, 36 f.)

Whenever we think or speak or even simply live (act, have aims and interests), we use logical forms. They rule our thoughts and beliefs, and dominate our actions and interactions.

We see then that the expressions "*das Logische*" and "logical nature" refer to logic as an objective fact, independent from human decision: the former denotes the natural field in which logical forms emerge; the latter expresses the natural and "unconscious" activity of using these forms. Now Hegel says that our "task" is to focus attention upon the forms of thought, making them the object of inquiry: they are used unconsciously, and we have to bring them into consciousness. This enterprise is what Hegel calls "*die Logik*", the theory or discipline that isolates and fixes the forms of valid inferences, "extracting them" from human language and life.

Forms are hence for Hegel objective occurrences, belonging to the domain of *das Logische* (human thought), and "logic" is the theory or the discipline that isolates and fixes them. Yet, the connection between *das Logische* and *die Logik* is not so immediate and uncontroversial. In the *Lectures on Logic and Metaphysics* (1817) Hegel writes:

> Logic is for us a natural metaphysics. Everyone who thinks has it. Natural logic does not always follow the rules which are established in the logic as theory; these rules often tread down natural logic. (Hegel 1992, 8)

Logic is natural. More specifically, it is a "natural metaphysics", as we will see better later. Thinking means to have "a logic" (an order of thought). But what is to be stressed is that this natural activity does not follow all and only the rules established by *die Logik*. Logical rules as fixed by *die Logik* sometimes, indeed often, "tread down" the naturalness of thought. "Natural logic" so parts company with "logic as theory".[2]

That our natural way of thinking is, in many senses, not strictly "logical", and that our reasoning is often ruled by "cognitive illusions" is for us quite ob-

---

[2] In the *Jäsche Logik* (Kant 1996, vol. 2, 439) Kant recalls that the distinction between natural logic (*natürliche Logik*) and scientific logic (*wissenschaftliche Logik*) is a common one in his times and is also known as the distinction between *logica naturalis* and *logica scholastica*. However, the true logic is for Kant only the scientific one. While in Kant natural logic is excluded from the scientific consideration, in Hegel it is an essential component of logic as a science.

vious.³ But Hegel has an opposite view: for Hegel what is wrong is, most frequently, logic, rather than the "natural" way of thinking. He stresses that the logical rules established by *die Logik* might be, and in fact often are, wrong, with respect to natural thought. This is precisely the dialectically relevant situation that introduces Hegel's particular criticism of "logic" as an apparatus, a fixed institutional discipline as it was practised in his times.⁴

Hegel's critique is linked to the "objective" conception of logic that he typically favours. As Nuzzo (1997, 47 ff.) has noted, the distinction between "natural logic" and "logical theory" implies on the part of Hegel a "non-instrumental" view of logic. For Hegel it is not the case that we *use* or *have* logical rules and forms. They are not an *instrument* to assess validity or truth. Rather, logic *has* and *uses* us, and all what we can do is to follow it, reconstructing the objective logical behaviour of thoughts and language.⁵

From this perspective, the reason why we acknowledge some logical constraint also becomes clear. The normative action of logical forms, as facts belonging to the realm of *das Logische*, is totally independent from our decision and arbitrary choice. In a sense, one might say that *das Logische* has an unquestionable primacy over *die Logik*. Natural inference relations present a necessity of their own, a necessity that is stronger than the supposed validity of rules fixed by logic as theory, and that has the right to be acknowledged as such. The naturalness of logical forms is what justifies the logical constraint, the constraint that forms exert on our thinking and believing. But it is this same naturalness of logic that grounds and explains Hegel's criticism of traditional logic.

There are many examples of the "failures" of traditional logical rules in Hegel's texts.⁶ One of the most vivid ones, presented in the short article "Who thinks abstractly?" (1807), is the anecdote of a prosecution. Common people, when a lady claims that a murderer who is brought to the place of execution "is handsome", are shocked and remark: "how can one think so wickedly and call a murderer handsome?". Since we are normally committed to the (logico-

---

3 See on the errors of natural reasoning the classical account of Wason/Johnson-Laird 1972.
4 Importantly, Hegel also criticised the philosophical habit of his times, particularly diffused among the romantics, to consider the study of logic as superfluous, and to reduce logic to an account of natural logic and a mere psychological consideration of one's thinking activity. See Krohn 1972, 56.
5 Nuzzo underlines this point in 1997, 47 ff. I agree, but I also stress the normative function of logic as theory on "the logical".
6 See for instance Hegel Werke 2, 575–581, Hegel Werke 18, 526–538 as well as Ficara 2013, 35–52, n. 1.

metaphysical) view that a same subject cannot have opposite properties, we can conclude that those who call a criminal (a bad person) "good" (intelligent, handsome) "think wickedly". Common people, in the example, represent the normative instance of logic as theory, and the lady's remark its violation, yet a violation that is, as Hegel shows, absolutely legitimate.[7]

In this regard, another distinction deserves to be mentioned, the one between *Verstandeslogik* and *Vernunftlogik*.

1.2 Two logics?

It is fairly evident that there are different ways of dealing with inconsistencies between "logic as theory" and "natural logic". If we stick to the validity of the rules, we have "intellectual" or "finite logic" (*Verstandeslogik*); if we question the validity of the rules, we have instead "speculative" or "rational logic" (which we may call, for symmetry, *Vernunftlogik*).[8]

With the expression "*Verstandeslogik*" Hegel generally refers to the traditional Aristotelian logic of his times, the theory of judgements, concepts and syllogisms, presented, among others, in Kant's *Jäsche Logik*. "*Vernunftlogik*", the logic of reason, *speculative* or also *dialectical* logic, is the logical enterprise in the specifically Hegelian sense. It involves a critical consideration of *Verstandeslogik*, that is, of the basic logical concepts and forms (among others: the concept of sentence, of concept, of contradiction, as well as the principle of excluded middle, the principle of identity and the principle of non-contradiction).[9]

Now it is common opinion that Hegel's view on intellectual logic is irretrievably critical.[10] However, Hegel's view is more complex, and cannot be reduced to

---

[7] On Hegel's standpoint regarding common sense see Vieweg 2007, 111 ff. Vieweg remarks that Hegel did not primarily criticise common sense, but rather the philosophies of common sense, according to which common sense is an unquestionable normative instance.
[8] The term is not used by Hegel as such, but Hegel employs different similar terms (*Logik als Wissenschaft der Vernunft* as well as the formulations *Vernunftschluss*, *Vernunfterkenntnis* etc.).
[9] In passing, we may note that, at a very preliminary level, *Vernunftlogik* is also an examination of the adequacy relation between *Verstandeslogik* and metaphysics (see here Chapter 2).
[10] For an interpretation of Hegel's critique of intellectual logic (the formal logic of his times) as a flat out rejection of formal logic see the *Historisches Wörterbuch der Philosophie*: Ritter/Gründer/Gabriel (eds.) 1971 ff., vol. 5, 358. In Adorno 2010 the link between *Vernunftlogik* and *Verstandeslogik*, the first called *Dialektik*, is reconstructed in terms very similar to mine. Adorno 2010, 304 underlines that "dialectics (*Vernunflogik*) presupposes the validity of the logical (*verstandes-logische*) laws and yet must go beyond them". Dialectics is the attempt to "escape from the prison of *Verstandeslogik*" not by fleeing to a pre-logical dimension but "by bringing [*Ver*-

a simple rejection of the intellectual approach to logic. As will be made clear in Part II, Hegel, though pointing out the limits of *Verstandeslogik*, also praises it, considering it the necessary basis of "the logic of reason". He writes:

> [The logic of] reason is nothing without [the logic of the] intellect, the [logic of the] intellect is still something without [the logic of] reason. (Hegel Werke 2, 551)

We can thus say that *Verstandeslogik* is the necessary – though not sufficient – condition of *Vernunftlogik*, insofar as the latter cannot survive without the former. Intellectual logic is the primary condition: it is the descriptive activity of grasping the natural activity of forms, and it also exerts a normative constraint on natural logic. Rational (dialectical) logic, in contrast, is the critical, normative reflection on the same intellectual logic. Hence rational logic ultimately justifies the application of the forms captured by the intellect to the effective and natural activity of reasoning. In this sense, *Vernunftlogik* is, in Kantian terms, the "tribunal" which intellectual logic has always to come before.

The distinction between the two "logics" in Hegel's account parallels Kant's distinction between *Verstand* and *Vernunft*. The distinction is of the greatest importance for Hegel, to the point that he places it at the centre of his own view about philosophy and philosophical rationality.[11] Hegel keeps the Kantian terminology, deepening it and changing, in some respect, Kant's conceptual framework. More specifically, differently from Kant Hegel postulates the idea of a rational logic (*Vernuftlogik*), which contains the critique of the forms established by logic itself (as *Verstandeslogik*). Hence Hegel also stresses[12] that it is one and the same discipline (namely logic as a science, the science of logic) that individuates the forms and conceptual determinations and at the same time criticises, or revises, them. I will limit myself here to hint at the elements of continuity between the two philosophers, leaving the question of the difference between them to

---

standeslogik] to reflect about its own insufficiency" (Adorno 2010, 306). Similarly, according to Stekeler-Weithofer 1992, 8 Hegel's logic is the general method of reflection about a conventional praxis including the higher order speculative reflection on the possibility conditions of thinking knowing and reasoning. Logic is not merely presupposed as a "general doctrine of thought" (*allgemeine Denklehre*), it is also criticised and analysed in its problematic nature. Hegel's logic is thus for Stekeler-Weithofer 1992, 9 a "general doctrine of the method of reflexive thought" (*allgemeine Methodenlehre des reflektierenden Denkens*).

11 Starting from his early writings (cf. Hegel Werke 2, in particular 20 ff., 305 ff. and 551) Hegel adopts this Kantian distinction. On *Verstand* and *Vernunft* in Kant and Hegel see Fulda/Horstmann (eds.) 1994, in particular 235–286.

12 See for instance Hegel Werke 6, 287/Hegel 1969, 611.

later considerations (Part II).¹³ This will allow us to better understand Hegel's view on the way in which *die Logik* must behave with respect to *das Logische*, or, more generally, how logical theory must grasp logical nature, without betraying it.

Kant uses *Verstand* and *Vernunft* both as synonyms for "thought", and the same holds for Hegel. In both Kant and Hegel the two terms define two different ways of thinking. *Verstand* is the faculty of judgements and partial determination, while *Vernunft* is the faculty of inferences and complete determination. The first is the faculty that Kant studies in the *Transcendental Analytic*, which analyses the *a priori* elements of our thought, identifying their common features, distinguishing them from each other and fixing them. The latter is the main subject matter of the *Transcendental Dialectic*, whose concern is how the forms and concepts isolated in the *Analytic* are applied to themselves. In the *Transcendental Analytic* the forms of judgements (which are studied in the logic manuals of Kant's and Hegel's times), as well as the corresponding categories (the basic forms of our knowledge of objects) are isolated and considered insofar as they are the very possibility conditions of objective knowledge. The *Transcendental Dialectic* deals with the same forms and shows how, if applied to themselves and not to the manifold given in the intuition, they give rise to paralogisms and antinomies. The *Transcendental Dialectic* is hence both *Logik des Scheins* (logic of illusion), insofar as the application of forms to themselves generates antinomies, and *Kritik des logischen Scheins* (critique of logical illusion), insofar as it shows how the illusions and mistakes generated by reason can be avoided.¹⁴

Thus for Kant as well as for Hegel "reason" is dependent on the intellect, and on the forms isolated by it. At the same time reason is more general than the intellect, insofar as it reflects upon intellectual forms and rules, discussing their application to the real contents of thought. For both Kant and Hegel reason is dialectical, while the intellect is analytical. *Verstand* thus separates and isolates the elements of our thought (the forms and concepts), while *Vernunft* re-

---

13 That dialectical, speculative philosophy tries to complete the project initiated by transcendental philosophy is stressed by many authors, among others Gadamer 1976 and Demmerling 1992, 67. For Demmerling the forms of thought are the specific research field of both dialectic and transcendental philosophy (see 79f.). On the importance of Kant for Hegel's philosophical project in the *Phenomenology of Spirit* see among others Wiehl 1966, 103ff. and Gadamer 1976, 35ff. On Hegel's interpretation and transformation of Kant's views on the antinomies see Düsing 2012, 93–114. On Hegel's interpretation of Kant see also Verra (ed.) 1981, Baptist 1986, Brinkmann 1994, 57–68, Engelhard 2007, 150–170, Sedgwick 2012.

14 Evidently, the two conceptions imply radically different attitudes toward antinomies, as we will see later, in Parts II and IV.

flects upon them, applies them to themselves, and in so doing generates irreducible and inevitable contradictions (antinomies).

We can see then that for both thinkers "intellectual logic" fixes rules and forms, while "rational logic" implies the critical reflection on those rules and forms, as well as the explanation of their relations to each other and to the totality of thought. The classical and generally acknowledged difference between the two accounts is due to the fact that Kant separates the critical and completely determining activity of reason from truth. For him, the formal criterion of truth is the Law on Non-Contradiction, while reason, in its attempt at fully determining things, encounters contradictions, and thus cannot preserve truth. For Hegel, in contrast, rational thought is driven by truth. As a matter of fact, we will see that truth for Hegel is complete and sceptical, so rational thought is complete and sceptical. Moreover, we will see that for Hegel both scepticism and completeness require contradiction in a significant way.

For now it is important to stress another difference. As intellect and reason are necessary parts of the human process of knowing and believing, so *Verstandeslogik* and *Vernunftlogik* are, for Hegel, necessary and related parts of *one single enterprise*. The former is the theory that fixes and isolates the forms "sunk" within natural language, while the latter is the critical analysis of these forms. Such an analysis leads, in some cases, to questioning the validity of forms established by *Verstandeslogik*.

In this respect, as we will see, the fault of traditional logic according to Hegel is that it tends to be developed only in terms of intellectual logic, so failing in its duties towards *das Logische*, and its rootedness in the reality of things – that is, in its duties towards truth. What thus now needs to be treated is the rootedness of logic in reality, hence the "metaphysical" commitment of logic in Hegel's view.

2 What does it mean to say that "logic coincides with metaphysics"?

In a famous passage of the *Enciclopaedia* Hegel writes: "Logic [...] coincides with metaphysics, the science of things set and held in thoughts" (Hegel Werke 8, 81/ Hegel 1991, 56). The close relation between logic and metaphysics is a key point in Hegel's philosophy, as well as an especially controversial and hardly understandable one, from a contemporary point of view.[15] One might even think

---

[15] Interestingly, the theme "Hegel and metaphysics" is origin of opposite interpretations. For some authors Hegel is a metaphysician (see, among the most recent works, the essays collected in De Laurentiis 2016). For others (see among the most recent contributions Jäschke 2012, 11–22) his philosophy is anti-metaphysical. For Stekeler-Weithofer 1992, 68 Hegel's talk about the relation between logic and metaphysics means that the two terms "logic" and "metaphysics" are "interchangeable titles for the conceptual analysis of rational thought". This can be taken to imply the idea of the superfluity of metaphysics (as the inquiry into what there is and its nature) in Hegel's philosophy. However, Stekeler-Weithofer's position in 1992 is more complex, and cannot be read as *tout-court* anti-metaphysical. In the Anglo-American bibliography the so-called "non-metaphysical" interpretation, defended, among others, by Pippin 1989 and Pinkard 1966, 13–20 (but see also Engelhardt/Pinkard (ed.) 1994) arose in opposition to traditional so called "metaphysical" interpretations of Hegel's philosophy as spiritualistic idealism or Platonism (see Stern's reconstruction in 1996). Pippin, Pinkard and Stern reject this metaphysical reading stressing the transcendental, Kantian and critical origins of Hegel's philosophy, whereby "transcendental" is taken to qualify a theory about thought and science, rather than about being or the structure of reality. I agree with the interpretations that highlight Hegel's transcendentalism. However, in 2006 I also showed how the transcendental and critical perspective, in Kant himself, is not to be read in anti-metaphysical terms. Evidently, the argument is settled as soon as one clarifies the meaning of "metaphysics" and "transcendental philosophy". For an analysis of the concepts of ontology, metaphysics and transcendental philosophy in Kant see Ficara 2006. For a reconstruction of the Anglo-American debate on Hegel and metaphysics see Beiser 1993, 1–24, Stern 1996, 206–225, De Boer 2011, 77–87, Ficara 2011, 400–405, De Laurentiis 2016, Zambrana 2017, 292f., Kreines 2017, 331f., Lebanidze 2019. If by "metaphysics" is meant: mysticism, dogmatism, or a specific worldview, then Hegel's logic (and philosophy) is not metaphysics. If by "metaphysics" is meant a theory about the most general structures of reality, then it is. Koch 2014, 238 specifies that Hegel's logic is not metaphysics if by "metaphysics" we mean a specific and non revisable theory about how things stand, but it is a metaphysics as "theory of the logical space", whereby Hegel held that the logical space, which Plato or David Lewis conceived in static terms, evolves (Koch 2014, 304ff.). In 2018, 220 Stekeler-Weithofer argues for a fundamental identity, in Hegel's logic, of logical conceptual analysis [*logische Begriffsanalyse*] and ontological analysis of forms [*ontologische Formenanalyse*]. That Hegel's logic is a consideration of the most fundamental structures of thought, which coincide with the most fundamental determinations of being is defended by Houlgate 2018, 146. In his most recent works, Pippin (2016, Chapter 7; 2017, 199–218; 2019) argues for the view that Hegel's reflection on the relation

that, in this regard, Hegel's notion of "logic" is in no way comparable to our modern conception, which is typically informed by formalism and ontological neutrality.

2.1 Metaphysics as a part of logic

In a letter to his friend Niethammer of October 23, 1812, Hegel observes that metaphysics is "a science about which one is nowadays accustomed to some embarrassment" (Hegel Werke 4, 406). Among the philosophical disciplines that are taught in the Gymnasium, "metaphysics seems to go away empty-handed, as psychology and logic have taken its place" (Hegel Werke 4, 406). At the same time, metaphysics, Hegel writes, is "completely maintained" within logic.

> According to my view, the metaphysical [*das Metaphysische*] in any case falls entirely within the logical [*das Logische*]. Here I can cite Kant as my precedent and authority. His critique reduces metaphysics as it has existed until now to a consideration of the understanding and reason. Logic can thus in the Kantian sense be understood so that, beyond the usual content of so-called general logic, what he calls transcendental logic is bound up with it and set out prior to it. In point of content I mean the doctrine of categories, or reflective concepts, and then of the concepts of reason: analytic and dialectic. These objective thought forms constitute an independent content [corresponding to] the role of the Aristotelian *Categories* [*organon de categoriis*] or the former ontology. Further, they are independent of one's metaphysical system. They occur in transcendental idealism as much as in dogmatism. The latter calls them determinations of being [*Entium*], while the former calls them determinations of the understanding. (Hegel Werke 4, 406 f.)

Hegel points out that the field of metaphysics (*das Metaphysische*) falls entirely within the field of logic (*das Logische*) and that, in this respect, his conception follows the Kantian one. In Kant the field of Aristotelian and pre-Kantian ontology,[16] which roughly corresponds to the subject matter of Aristotle's *Categories*,

---

between logic and metaphysics is fundamentally coherent with the ancient (in particular Aristotelian) conception (see Pippin 2016, 168, 171, 181, 187). I have defended a similar view in Ficara 2014c, 245–256. Many classical works (among them Gadamer 1976, Verra 2007, Düsing 2012, Riedel (ed.) 1990a) argue for the substantial continuity between the Hegelian approach and ancient philosophy on this point. As Gadamer 1987, 93 recalls, Heidegger used to say in this context: "Hegel, der radikalste Grieche (Hegel, the most radical Greek [philosopher])". For a critical consideration of Pippin 2019 see Baumann 2019, 1256–1260.

16 The term "ontology" appears for the first time in Goclenius' *Lexicon philosophicum* (Goclenius 1613) and means *philosophia de ente*. Following this use, Wolff 1730, § 1 calls "ontology" Aristotle's *philosophia prima*, identifying it with the science of being qua being (*scientia entis in genere, quatenus ens est*) and of all principles of human knowledge. Baumgarten 1739, § 4 de-is the topic of transcendental (analytical and dialectical) logic and is dealt together with (Hegel says: set prior to) the content of formal logic (as theory about the forms of concepts, judgements and inferences). Hegel states that formal logic and metaphysics (intended as category theory) belong to each other because their subject matter is independent of one's metaphysical system (be it, for example, transcendental idealism, or dogmatism). Interestingly Hegel distinguishes between "metaphysical system" (*metaphysisches System*) and "the metaphysical" (*das Metaphysische*). We clearly see that the expression *das Metaphysische* recalls the expression *das Logische*. "Metaphysical system" means here a specific theory about what there is and about its nature, while *das Metaphysische* refers to the most general net of concepts and forms of our thought about reality, exactly like *das Logische* stands for the effective occurring of logical facts in our activity of thinking and believing.

In this regard, we can fix a first (Kantian) meaning of the expression "logic coincides with metaphysics", namely that *the subject matter of earlier, i.e. pre-Kantian, ontology (as the theory about the most general structures of our thought about reality) should be dealt with within logic (as theory about the most general forms of thought)*.[17]

2.2 Logic and the objectivity of thought

In the Introduction to the *Science of Logic* Hegel considers the question of the relation between logic and metaphysics in the context of an analysis of logic's subject matter, and of a critique of merely formal or subjective conceptions of

---

fines ontology as "the science of the most general and abstract predicates of anything" (see on the history of ontology among others Ferrater Mora 1963, 36–47, here 36). In short, Kant's contribution to the ontological tradition consists in underlining that the first "principles and predicates of being" are concepts and structures of *thought* (and only through thought of being). Consequently, Hegel is right when he states that in Kant the field of ontology "falls entirely within the field of logic". It is precisely for this reason that Kant, famously, proposes to replace the term "ontology" with "analytic of the pure understanding" (*Analytik des reinen Verstandes*), which is a part of (transcendental) logic. On Kant's view on the relation between ontology, logic, and transcendental philosophy see Ficara 2006.

17 In passing, we may note that the definitions of "ontology" and "metaphysics" presupposed in Hegel's account are partially consistent with what we may intend now with the two terms, and especially in recent analytic philosophy, where "ontology" is the study of "what there is" and "metaphysics" is the study of the nature of what there is (see Varzi 2011, 407). For an overview about the meaning of metaphysics in both analytical philosophy and the history of philosophy see d'Agostini 2008b, 244–270.

thought. Hegel notes that ancient philosophy had a "higher" conception of thought than the one typical of modern philosophy. According to this conception, real is only what is graspable through thought, and thought is the very ground on which we can grasp something as existent. According to Anaxagoras, for instance, *nous* is "the principle of the world, and the essence of the world is to be defined as thought" (Hegel Werke 5, 44/Hegel 1969, 50). According to Plato, "something has reality only in its concept" (Hegel Werke 5, 44/Hegel 1969, 50). Hegel also remarks that this is an *objective*, and not a psychological or subjective, conception of thought, a conception that is also "already present" in our general and common idea of logic:

> [O]ne can appeal to the conceptions of ordinary logic itself; for it is assumed, for example, that the determinations contained in definitions do not belong only to the knower, but are determinations of the object, constituting its innermost essence and its very own nature. Or, if from given determinations others are inferred, it is assumed that what is inferred is not something external and alien to the object, but rather that it belongs to the object itself, that to the thought there is a correspondent being. (Hegel Werke 5, 44/Hegel 1969, 50 f.)

In this sense, as I have suggested, logical rules are not arbitrary, for Hegel: they are the very expression of the structure of reality. Similarly, in the *Vorbegriff* of the *Encyclopaedia Logic*, Hegel writes:

> If thought tries to form a concept of things, this concept (as well as sentences and arguments) cannot be composed of parts and relations which are alien and irrelevant to the things. (Hegel Werke 8, 81/Hegel 1991, 56)

Thus Hegel states that the inferential forms fixed by logic are supposed to "belong to the object", insofar as the correspondence of being and thought is given as such, in the very same structure of reality and human thought.

In the *Vorbegriff* of the *Encyclopaedia Logic* Hegel also calls the logical forms "objective thoughts".

> With these explanations and qualifications, thoughts may be called *objective thoughts* – among which are also to be included the forms which are more especially discussed in the common logic, where they are usually treated as merely forms of conscious thought. *Logic therefore coincides with metaphysics, the science of things set and held in thoughts* – thoughts accredited able to express the essential reality of things. (Hegel Werke 8, 81/Hegel 1991, 56)

Hegel specifies here that the field studied by logic is objective thought,[18] i.e. "thought accredited to express the essential reality of things". In this sense logic coincides with metaphysics simply because logical forms capture the formal aspects of our thinking things, the formal aspects of things as they are grasped by and held in thought.

It is important to stress that Hegel's theory of "objective thought" does not imply an "epistemic" conception of logic and metaphysics, and does not imply a subordination of metaphysics (and logic) to epistemology either. Hegel does not say that reality as such is contained in thought, or has only a noetic, cognitive nature. He rather says that *die Logik* (logic as theory) includes *die Metaphysik* (metaphysics as theory or discipline), since *das Logische* (the domain of logical forms in natural thought) contains the structure of things as we know them: *das Metaphysische*.

Accordingly, Hegel criticises the subjectivistic and formalistic approach to logic, defended, among others, by Kant,[19] in virtue of which logical forms are "merely forms of conscious thought [nur Formen des bewußten Denkens]". This would actually mean that logical rules are taken to be mere expressions of how an epistemic subject should think and not (also) as expressions of how things stand.[20] In criticising this approach, Hegel defends an Aristotelian position about the correspondence between *logos* and *on*. As a matter of fact, we will see that Hegel's conception of logic is entirely oriented by the ancient idea of truth. For now, we may sum up isolating a second meaning of the statement "logic coincides with metaphysics" namely:

the forms of thought studied by logic express the same structure of things.

---

18 The Hegelian expression "objective thought(s)" does imply a kind of Platonism about concepts and forms, as well as a critique of a psychologistic view of thought. As such it is close to the Fregean idea of objective thought. See on Frege's and Hegel's objective thought d'Agostini 2003, 59–94. On objective thought in Hegel see Halbig 2002. What Hegel stresses in this passage is, however, not primarily the fact that logical laws and forms *are* in themselves something real, but that they *express (correspond to)* reality. In this respect, the focus here is rather on the Aristotelian idea of the coincidence between logic and metaphysics as based on the necessary link between (forms of) thought and reality. On the Aristotelian heritage of Hegel's account of objective thought see Ferrarin 2001, Chapter 4.
19 See on this Part II.
20 For Quante 2018, 277 Hegel's program in the *Science of Logic* is developing a "theory of absolute subjectivity" which unifies the "ontology of the logic of being" [*seinslogische Ontologie*] of the *Science of Logic*'s first part with the "ontology of self-consciousness" [*Ontologie des Selbstbewusstseins*] of the *Science of Logic*'s second part.

2.3 Logic and metaphysics. From nature to theory and back

The thesis of the inextricable connection between logic and metaphysics is the mark of Hegel's Aristotelianism[21]. That the logical constraint comes from reality is a typically Aristotelian view, well expressed in Book IV Chapter 4 of the *Metaphysics*. As we have seen, Hegel thoroughly accepts the idea. Not only that, he finds it again in Kant's conception. He thus provides for a conciliation between the Aristotelian and the Kantian vision of metaphysics, showing that there is no true incompatibility between them.

In conciliating Aristotle and Kant, as well as in recovering the ancient connection between logic and metaphysics, what we have seen about natural logic and natural metaphysics plays a special role. As we have seen, Hegel distinguishes between logic as theory (*die Logik*) and logic as natural logic (*das Logische*) and conceives the latter as "natural metaphysics":

> Logic is for us a *natural metaphysics*. Everyone who thinks has it. *Natural logic* does not always follow the rules which are established in the *logic as theory*; these rules often tread down *natural logic*. (Hegel 1992, 8. Italics are mine)

I have already suggested that in the same way as we have a natural and a theoretical level in logic we should have, in principle, a natural and a theoretical level in metaphysics. Logical patterns are present in our natural thinking/writing/speaking, whether we like it or not; the task of logic as theory is to make these patterns explicit. Similarly, metaphysical conceptions are present in our everyday thinking – "human beings are born metaphysicians" (Hegel Werke 8, 207/Hegel 1991, 156) and the task should be to make these conceptions the object of inquiry in our metaphysics as theory. Thus, we have an identical relation between a *natural* and a *theoretical* level in both disciplines.

The natural level of human thought contains both the forms of things (natural metaphysics) and the inferential forms human beings deploy in their natu-

---

[21] An interesting piece of evidence is to be found in the debates about Hegel's dialectics in the so called *Berliner Aristotelianism*. See on this Ficara 2015, 39–55. The Aristotelian Trendelenburg, despite his critique of Hegel's logic, held a view on the link between logic and metaphysics that is perfectly consistent with the one sketched above. See on logic and metaphysics in Trendelenburg Peckhaus 2013, 283–296, as well as Gabriel 2007, 237–240 and Peckhaus 2007, 241–255. On Hegel, Trendelenburg and the "logical question" [*die logische Frage*] see Gerhard 2015. On Hegel's interpretation of (and fundamental continuity with) Aristotle's logic see Mignucci 1995, 29–50. On the compatibility between Hegelian and Aristotelian logic see also Redding 2007. More generally, on Hegel and Aristotle see Verra 2007, 349–370, Ferrarin 2001 and Düsing 2012, 131 ff.

ral exercise of thought (natural logic). The natural coincidence of logic and metaphysics is what informs *Vernunftlogik*. The task of *Vernunftlogik* in its critique of *Verstandeslogik* is oriented by truth, that is, by the "coincidence" between *forms of things* and *forms of reasoning*.[22]

Significantly, Hegel says that "[natural] logic is for us a natural metaphysics". The reason of this is that metaphysics, for Hegel, is simply the net of thought determinations that we use in our natural thinking and speaking about reality.

This does not mean that "metaphysics" does not exist as such. At the natural level, there is no separation between the determinations of thought (inferential forms, logical principles etc.) and the determinations of reality (quantity, quality, being etc.). They are all structures that orient our life and reasoning, and we use them without even being aware of them. We might suppose, instead, that at the theoretical level, when these structures become an object of inquiry, the distinction between metaphysical and logical forms takes place, and they become domains of different disciplines. Interestingly enough, the Hegelian conception entails that the natural-level interplay between logic and metaphysics also survives at the theoretical level. Hegel develops his *Wissenschaft der Logik* accordingly, as an analysis of determinations that are traditionally taken to be determinations of being (like "quantity", "quality", "measure") and determinations of thought (like "concept", "judgement", "syllogism"), alongside an inquiry about their relations to each other.

On this basis, we can fix a further aspect clarifying the "coincidence" of logic and metaphysics: *logic and metaphysics, though strictly connected at the natural level, are distinguished at the theoretical level, and their natural connection is restored in the Hegelian idea of a 'Science of Logic'*. In other words, the specific concern of a science of logic intended as *Vernunftlogik* is to restore, critically and normatively, the natural identity of logic and metaphysics: in a sense, it should put the logical forms devised by *Verstandeslogik* before the tribunal of metaphysics, i.e. of truth. This should happen because it is our natural metaphysics (our natural conception of things) that is *normative* for logic, just as, for Aristotle, it is the same nature of being, as substance, that guides our reasoning and speaking.

---

[22] On the link between *Vernunftlogik* and truth see III.

2.4 From natural logic to philosophical logic

These are, sketched in very general terms, the basic ideas that inspire Hegel's conciliation of Aristotle and Kant, and his reinstating logic upon its "naturally metaphysical" grounds. But the whole picture makes sense because the human enterprise that deals with the ontological naturalness of forms (that is: philosophy) has and should have a sceptical, critical attitude. Logic and metaphysics are one and the same just insofar as they are dialectical, i.e. philosophical, critical and self-critical disciplines. In the *Logic* of the *Encyclopaedia* Hegel writes:

> [H]uman beings are born metaphysicians. And the question is only if the metaphysics one uses is of the right kind, if we keep with univocal and fixed intellectual determinations as the basis of our theoretical and practical activity, instead of keeping with concrete, logical ideas. (Hegel Werke 8, 207/Hegel 1991, 156)

Similarly, in the *Naturphilosophie* of the *Encyclopaedia* we read:

> What distinguishes the philosophy of nature from physics is the kind of metaphysics it adopts. As a matter of fact, metaphysics is nothing else than the range of thought determinations, the network in which we bring every matter and through which we make it understandable. Every educated mind has its metaphysics, the instinctual thought, the absolute power in us over which we become master when we make it the object of our thought. (Hegel Werke 9, 20)

The metaphysical views orienting, as Hegel says, "our theoretical and practical activity" can be problematic: they force us to think in certain ways and to do certain things. When Hegel points out that the main problem is therefore not if we have a metaphysics or not, but rather if our metaphysics is wide and flexible enough, he also refers to the risks of holding to a too rigid network of thought forms. Therefore he says that there is only one right metaphysics, and this is the "concrete, logical idea". By "concrete logical idea" he means *Vernunftlogik*, which is one and the same as dialectical logic, typically defined by Hegel as the logic of "concreteness". There is no need to deepen the meaning of dialectic here.[23] For now, we can see that the coincidence of forms of things and forms of thought is theoretically guaranteed because "the science of forms" is performed in a *philosophical* – that is complete, and therefore critical, dialectical – way.

In this sense we may suitably say that Hegel's logic – as we will see in the last chapter of this part – is what we can define nowadays as a *philosophical*

---

23 For an analysis of the concept in Hegel, as well as of its history, see Part IV below.

logic. Interestingly, Hegel himself (in the *Philosophy of Right*) introduces the expression "philosophical logic" to refer to his own idea of logic.

2.5 Philosophical logic as conceptual logic

The expression "philosophical logic" occurs in fact for the first time in Hegel's work. In § 2 of the *Philosophy of Right* Hegel explains that his exposition of the concept of right follows the "scientific method of philosophy", the one presented in the "philosophical logic" (Hegel Werke 7, 32). In this light, it is fairly clear that his idea of philosophical logic was something like *"logic seen form a philosophical perspective"*, which for Hegel means from a *scientific*, that is, *complete* and *speculative*, perspective.

Hegel then sums up the main traits of the "scientific method in the philosophical logic", saying that it consists in focusing on the *concept*, which is called here both "the form" and "the truth" of the problem which is to be analysed, and stating that this also implies examining common and ordinary thoughts. The concept is "necessary" – insofar as it coincides with the truth about the object at stake, the concept of right – and one should, Hegel writes:

> [L]ook at what, in representations and language, corresponds to it. However, as far as the form is concerned [...] there has to be a difference between how this concept is in itself in its truth and how it is in representation. If the representation is not false according to the content, then it is possible to show the concept as contained in it and present in its essence, that is, the representation can be brought to the form of the concept. But the representation is not criterion of the concept, which is in itself necessary and true. The representation has rather to draw its truth from the concept, to rectify and know itself through it. (Hegel Werke 7, 32)

It is not my aim to elaborate here the meaning of the expression "the concept" in Hegel, or to explain the dialectical pattern of the scientific method hinted at by Hegel in this passage.[24] What is interesting to note at this point is that Hegel identifies the task of philosophical logic with "finding the concept", and "the concept" with both the *form* and the *truth* of the representation (our ordinary natural thought about the matter at stake). In Hegel's view, philosophical

---

24 For a detailed analysis see Part IV and V below.

logic itself is the operation of extracting the form (the concept) from common language and reasoning ("common representations").[25]

The task of the *Philosophy of Right* is the analysis of the concept of right, that is an explanation of what this concept truly means, such that a genuine evaluation of particular systems of laws and juridical institutions becomes possible. In order to achieve this, Hegel explains, one has also to look for what is *normally meant* by the concept of right: the concrete, common representation in natural language and thought. What we normally understand by "right" is different from the *scientific* (philosophical) concept of right: it does not have any scientific form yet. However, for Hegel it is possible to show that the scientific concept is already present in our preliminary conception, or, what is the same, the scientific concept can be extrapolated from our normal representation. This happens just because our normal way of thinking about it *contains elements of truth*. We have already seen this point, with reference to the "errors" of logic as theory. In turn, Hegel says, the criterion of truth cannot be the common representation, but it is to be found in the *fully developed* scientific concept of right. As soon as we have the completeness of the concept we have its form, and the truth about its content (what it is about).

In sum, as Gadamer 1976 clearly points out, what Hegel calls "das Logische" is the realm of conceptuality. Concepts are rules for the human expression of being and reality, and reason has the task of giving an account of them. Thus *die Logik* as *Vernunftlogik* has the task of giving an account of the conceptual realm, which is one and the same as the realm of norms and forms "sunk" within human thinking and reasoning (*das Logische*).[26] And the conceptual realm is the realm of thought about reality, i.e. what we claim is true or false. This means that

---

[25] I will show in what follows how the task of extracting the forms of thought is precisely the aim of what Russell (to whom nowadays is commonly retraced the meaning of the notion "philosophical logic") calls "philosophical logic".

[26] Merker 1996, 92 highlights that "form" and "concept" for Hegel overlap. Both are conceived in dynamic terms, and refer to the activity of "distinguishing what is identical and rendering identical what is different" (see also Encyclopaedia § 314). Also Stekeler-Weithofer 2005, Chapter 3 emphasises the connection between concept and form in Hegel, with special reference to its Platonic origins. Hegel's view on the conceptual realm (*der Begriff*) implies for Stekeler-Weithofer 1992, 40 the idea of an interplay between syntax and semantics. Logical forms are for Hegel "conceptual, i.e. syntacto-semantic forms". Accordingly, Hegel's logic is not, for Stekeler-Weithofer 1992, 40 fixation of forms (like Aristotle's syllogistic and Frege's logic) but meta-science, i.e. its aim is the critical analysis of logical forms. I stress that Hegel's logic is both fixation and critique, and that this happens because forms, for Hegel, are self-revising structures. For Hegel, it is impossible to revise forms without fixing them, and to fix them without revising them. On the revision of logic in Hegel see Ficara 2019, 59–72.

das Logische is intrinsically connected with *das Metaphysische*, and logic as theory is linked to metaphysics as theory. Insofar as logic as theory gives an account of the conceptual, i.e. of the forms and norms of thought about reality, it also gives an account of the forms of reality.

3 What kind of logic is Hegel's logic?

What I have tried to reconstruct so far is not Hegel's logic, but rather Hegel's general view about the notion of "logic". I have tried to stress those aspects of the view that might be most interesting for our contemporary convictions, and for the current ways in which logic is practised. Now I can move to a preliminary confrontation between Hegel's and contemporary views.

More specifically we have seen that, in Hegel's use, "logic" defines quite a wide enterprise, which encompasses the traditional (and still fairly standard) meaning of logic as "theory of valid inference" and the critical reflection on logic traditionally intended. Is this (can this be) consistent with what we intend – or what is generally intended – by "logic", nowadays?

Providing an answer to this question has a certain importance because there is a tendency, in the literature, to cut short with the matter, claiming that Hegel's notion of "logic" is in no way comparable to our notion, as he, with this word, ultimately intended something else. To quote only two among the common accounts of Hegel's logic within the history of logic, the *Historisches Wörterbuch der Philosophie* suggests that Hegel's view implies a "complete rejection of formal logic", and "its substitution with a dialectic which is the product of speculative metaphysics" (Ritter/Gründer/Gabriel (eds.) 1971 ff., vol. 5., 358). Similarly, Kneale and Kneale write that with Kant's transcendentalism "began the production of the curious mixture of metaphysics and epistemology which was presented as logic by Hegel and the other Idealists of the nineteenth century" (Kneale/Kneale 1962, 355).

In this regard, my strategy consists first in considering, very briefly, Hegel's peculiar position within the "transcendentalist" account of logic in Kant and Fichte. Second, I stress that Hegel's conception has a strong affinity with what can be intended as "philosophical logic" nowadays. Third, I hint at what Hegel's particular vision (informed by the notions of *das Logische* and *die Logik*, intellectual and rational logic, and "logic as metaphysics") can still tell us about the nature and tasks of logic.

3.1 Transcendental logic?

In the most widely shared meaning, "logic" is the discipline inquiring into "what follows from what"[27], i.e. the study of valid inference. This is the definition repeatedly suggested by Aristotle, for example in the *Prior Analytics*:

> A deduction is speech (*logos*) in which, certain things having been supposed, something different from those supposed results of necessity because of their being so. (Aristotle, *Prior Analytics* I, 2, 24b, 18–20 – translation from Barnes 1984)

In the transcendental and idealistic tradition this meaning is not substantially discussed, and it largely corresponds to what Hegel calls *Verstandeslogik* or also "usual logic [gewöhnliche Logik]" and Kant and Fichte call "formal logic".[28] As we have seen, Hegel stresses that *Verstandeslogik* is the necessary (but not sufficient) condition of *Vernunftlogik*. Similarly, Kant bases his transcendental (both analytical and dialectical) logic on the forms set up in formal logic. Fichte underlines that both his transcendental logic and formal logic have thought (forms of thought) as their object.[29]

---

[27] Quine 1986, vii quotes Tweedledee in Lewis Carroll's *Alice Through the Looking Glass:* "Contrariwise [...] if it was so, it might be; and if it were so, it would be: but as it isn't, it ain't. That's logic". Another definition is "logic as theory of formal systems". See for example Hodges 2007, 9 who distinguishes between two meanings of first order logic: "collection of closely related artificial languages" and "study of the rules of sound argument".

[28] Hegel preferentially uses the expression "gewöhnliche Logik", which means "common" or "usual" logic, and not primarily "formal logic". As we will see, this specifically Hegelian use has important philosophical implications. Interestingly, Kant introduces the expression "formal logic", but the expression entered into common use later (see Strube 1973).

[29] On the relation between transcendental and formal logic in Kant see among many others Barone 1957, Stuhlmann-Laeisz 1976 19 ff., Wolff 1984, 178–202. For further bibliographical references on the subject see also Ficara 2006, 153–155. On the meaning of "logic" in Fichte see Bertinetto (ed.) 2004, and in German Idealism in general Lejeune (ed.) 2013. For an in depth analysis of the differences between transcendental and speculative logic, and of Hegel's own interpretation of Kant's distinction between formal and transcendental logic see Nuzzo 2014, 257–273. Nuzzo (2014, 257) remarks that Hegel sees his own logic as a prosecution of the process Kant had started with his transcendental logic, but she also highlights the differences between the two positions. Hegel's position, for Nuzzo, converges with the Kantian one insofar as both philosophers link the task of logic (as transcendental viz. speculative) to the one of establishing the conditions of objective knowledge. The divergence concerns, for Nuzzo, the different conception of knowledge, and its relation to truth, held by the two thinkers. For an analysis of Hegel's interpretation of Kant's concept of truth in relation to logic in the *Subjective Logic* and a critical appraisal of Nuzzo's interpretation see part II. Taylor 1983, 299 ff. stresses the ontological character of both transcendental and dialectical logic, and interprets it as sign of the non-formal na-

Interestingly, while Kant and Fichte refer to their logical inquiry as "transcendental logic", Hegel labels his own inquiry as "logic", without further specification.[30] This suggests – in my view – that he gives voice to the idea of a complete "re-capture" of traditional formal logic within his speculative dialectical logic (*Vernunftlogik*). As a matter of fact, in Hegel's view there is no true duplication of the regime of logic. Logic is one discipline, including intellectual logic as its part, or phase or moment, within the speculative enterprise he calls "logic". In contrast, the Kantian and Fichtean use does suggest the view of transcendental idealistic logic as an *alternative*, with respect to common logic. In other words, Hegel's *Vernunftlogik* is not conceived in an oppositional or polemical way to the Aristotelian (formal) conception. It does imply a criticism of this conception (especially as developed in modern times). But, as we will see, the reason of the critique is merely the fact that traditional Aristotelian logic is *incomplete*.[31]

A second point worth noting is that in the transcendental tradition the term "logic" stands for a broader enterprise than the one designated by the term nowadays. It includes issues that we would rather think of as belonging to the philosophy of mind, or to epistemology, or to the philosophy of science, such as the analysis of the *a priori* elements of "thought" generally intended, or the foundations of science, or the nature of beliefs and knowledge.[32] In this sense, Kneale and Kneale's mention of the "curious mixture" may seem justified.

As we have seen, *die Logik* (logic as theory) refers for Hegel to a discipline that is more general than "logic" strictly intended (common logic). And yet, in

---

ture of Kant's and Hegel's logic. On the differences between transcendental and dialectical/speculative approach to formal logic see also Düsing 2012, 11 and 180. On Hegel's critique of Kant's formalism see Sedgwick 2012, Chapter 1.

30 The Hegelian expression *Wissenschaft der Logik* (science of logic), refers to the fact that, for Hegel, the aim of logic as a science consists in bringing into consciousness the essence of thought in its truth (natural logic and metaphysics), as well as its relation to *die Logik* (logic as theory). See on this Nuzzo 1997, 49. For Koch 2018, 44 f. in his *Science of Logic* Hegel develops an interplay between *Hintergrundlogik* (our pre-theoretical, everyday, implicit views about thought and being – which I, following Hegel, call natural logic and metaphysics) *Vordergrundtheorie* (thought as such [*das Denken als solches*] as the pure and "presuppositionless" thematic object of the *Science of Logic*) and *Hintegrundtheorie* (our reflection about the thematic object of the *Science of Logic*).

31 The question of Hegel's attitude toward formal logic is dealt with in part II. For an analysis of the link between Hegel and formal or "common" logic see Krohn 1972, Hanna 1986, 305–338, and, more recently, Redding 2014, 281–301.

32 This enlarged conception might be related to the Stoic view, see on this Hegel's interpretation of Stoic logic in Hegel Werke 19, 268 ff./Hegel 1892 ff., vol. 2, 249 ff. For a consideration of Hegel's interpretation of stoic logic see Redding 2014, 281–301.

Hegel's view "logic" does not properly denote a "mixture" of epistemology, metaphysics, and analysis of mind (thought). It is, instead, a discipline *well focused on what I have called "the logical fact"*, the *fact* of logical forms that "arise" in natural thought and language, and the *fact* of their normativity. *Vernunftlogik*, which is the genuine "logic as theory", thus contains *Verstandeslogik*, insofar as it is both the individuation of logical forms and their critical analysis. This critical-speculative account also involves, as we have seen, asking about the relations of logical forms to the forms or essences of things, so establishing the connection between logic and metaphysics. Not only that, it is also based on our natural activity of thought. All this does not produce a "mixture", rather: it promotes the *groundedness* of logic on logical facts, which are given by both the nature of things, and the nature of human thought about things.

Now the question is: is there, today, a meaning of "logic" corresponding to what Hegel meant by *Vernunftlogik*, that is to the enterprise that both fixes the forms of thought and critically reflects upon them, their relations to each other, and also their relations to reality and natural thought?[33]

3.2 Hegel's logic and contemporary conceptions of "philosophical logic"

What is reasonable to admit is that Hegel's logic can be conceived as what one ought to call nowadays "philosophical logic"[34]. There are disagreements and doubts about this expression, and it is frequently assimilated to (or not well distinguished from) "philosophy of logic".[35] But there are some positive aspects in current definitions that suggest a significant accordance with Hegel's view.

---

33 In 1976, 95 ff. Kambartel distinguishes between three aspects concerning the history and the meaning of "logic", the first is the link between logic and *lógos* (intended as rationality, rational confrontation among individuals and dialogue); the second is the relation between logic and forms, and the idea that logic deals with the form rather than with the content of thought; the third is the link between logic and truth (in this sense logic arises because it is possible to distinguish between true and false sentences). For Kambartel the relationship between dialectics and logic is to be retraced only to the first aspect.
34 Hegel, as we have seen, was very likely the first who used this expression. In the continental tradition it has been used with polemical intents, as opposed to "mathematical logic". What I am stressing here is that this polemical or contrastive connotation is not totally justified, since the analytical conception of philosophical logic, though developed in continuity with mathematical logic, is adaptable to Hegel's idea of logic. On the meaning of "logic" in the analytic and continental tradition see d'Agostini 2000.
35 Some authors explicitly identify them, see for example Jaquette (ed.) 2007, 2.

Extracting forms – The expression "philosophical logic" (PL) in the analytic tradition is traced back to Russell.[36] In *Our Knowledge of the External World* Russell writes:

> Take (say) the series of propositions "Socrates drank the hemlock," "Coleridge drank the hemlock," "Coleridge drank opium," "Coleridge ate opium." The form remains unchanged throughout this series, but all the constituents are altered. Thus form is not another constituent, but is the way the constituents are put together. It is forms, in this sense, that are the proper object of philosophical logic. It is obvious that the knowledge of logical forms is something quite different from knowledge of existing things. The form of "Socrates drank the hemlock" is not an existing thing like Socrates and the hemlock [...] some kind of knowledge of logical forms, though with most people is not explicit, is involved in all understanding of discourse. It is the business of philosophical logic to extract this knowledge from its concrete integuments, and to render it explicit and pure. (Russell 2009, 34 f.)

The Hegelian spirit of this quotation is outright clear.[37] Logical forms for Russell (as well as for Hegel) are always involved in our concrete talking with each other and understanding each other. They have "concrete integuments" (for Hegel they are even object of empirical analysis, as I discuss in Part II). Our talking and reasoning follows logical patterns, and this often happens implicitly, without any precise awareness on our part (Hegel stresses, as we have seen, the "unconscious" use of forms). The task of philosophical logic is then to "extract the knowledge about forms from its concrete integuments" (the same task of *die Logik*, in Hegel's sense), making the logical structure of our thinking explicit.

The idea of "natural logic", and of logical forms as (special kinds of) "linguistic facts" is hence at the basis of the preliminary way in which contemporary philosophy has conceived the notion of "philosophical logic". Following Russell, many contemporary authors (Sainsbury 2001, 1, Jaquette (ed.) 2007, 1, Cook 2009, 221) define philosophical logic as *the attempt to formalise natural language*, which might be performed by constructing mathematical models or more or less idealized languages. In any case, "formalisation" still means, ideally, what Russell calls "extracting" the forms that are entangled in our ways of speaking and thinking.

Logic for philosophy – Professedly inspired by Russell is another view endorsed among philosophers of logic. It is the view according to which "philo-

---

36 See Sainsbury 2001, Lowe 2013 among others.
37 On Russell's idealistic philosophical formation see Hylton 1990, 72 ff. and on the role of Hegel in early analytic philosophy Milkov 2020.

sophical logic" is, more specifically, a "logic for philosophy". This is well expressed by Mark Sainsbury:

> Russell coined the phrase 'philosophical logic' to describe a program in philosophy: that of tackling philosophical problems by formalising problematic sentences in what appeared to Russell to be the language of logic: the formal language of *Principia Mathematica*. (Sainsbury 2001, 1)

Dale Jaquette (2007, 1) also calls "philosophical logic" the discipline involving "applications of any recognised methods of logic to philosophical problems". And according to Roy T. Cook (2009, 221), philosophical logic "involves the use of formal systems as a tool for solving, or contributing to the solution, of philosophical problems". So stated, philosophical logic is "logic for philosophy", insofar as its focus is philosophy.[38]

That some conception of philosophical logic in this sense can also be found in Hegel has already been suggested.[39] More specifically, dialectic, as *Vernunftlogik*, is for Hegel the genuine logic to be practised in philosophy. So it is the genuine philosophical logic. But it should be stressed that in Hegel there is no trace of the idea, common to Russell and his commentators and followers, of the "application" to philosophy of a certain – separated – discipline. For Hegel logic *is* "philosophical" to the extent that it is a *complete* and hence *rational* discipline. In other words, non-philosophical logic is only a part of the enterprise of "logic as theory". The divergence between Hegel and Russell on this point is easily explained considering that the institutional creation of "logic" as a separate science, as well as the strict connection between logic and mathematics, were promoted, in these terms, by Russel himself, and would have been hardly conceivable in Hegel's times.

---

[38] Goble (ed.) 2001, 1 stresses this point, writing that philosophical logic "is philosophy that is logic, and logic that is philosophy. It is where philosophy and logic come together and become one. Philosophical logic [...] comprises the sorts of logic that hold greatest interest for philosophers. [It] develops formal systems and structures to be applied to the analysis of concepts and arguments that are central to philosophical inquiry. So for example such traditional philosophical concepts as necessity, knowledge, obligation, time and existence, not to mention reasoning itself, are usefully investigated through modal logic, epistemic logic, deontic logic, temporal logic".

[39] Croce 1906, Litt 1961, Günther 1978 claim that Hegel's logic is a logic of philosophy, i.e. an inquiry into the questions "what are philosophy's specific objects?" "what is philosophy's peculiar method"? For Litt 1961, 14 f. our thought proceeds in temporal succession, its objects are articulated in "parts, pieces, sections" that develop temporally. In philosophy each part is only conceivable together with the whole of which it is a part. Also Berti 2015 interprets dialectics as the logic of philosophy.

From logic to philosophy and back – Some authors have also claimed that the expression "philosophical logic" indicates a reciprocity between philosophy and logic, for reasons that might be linked to the two previously discussed hypotheses: the strictly Russellian (extracting forms from natural thought) and the derivatively Russellian (applying logical forms to the solution of philosophical problems). The relation of reciprocal foundation between logic and philosophy means first that the logical effort of making philosophical problems clear has implications for the same development of logic. Second, and more specifically, the attempt at applying formal methods to the analysis of philosophical problems implies that new forms are individuated, and new developments in formal logic are promoted.[40] Thus the (philosophical) field analysed by logic retroacts on the task of logical analysis (extracting forms). This is the reason why it is also a common view that philosophical logic is to be understood in non-classical terms, as a logic that challenges the classical canon of logic.[41]

The idea of philosophical logic as a study specifically engaged in proposing non-classical accounts of logic is, possibly, the contemporary conception of philosophical logic most closely related to Hegel's view (see also here IV and V). As it is well known, Hegel's philosophical logic has been (and can be) intended as "non-classical".[42] As we have seen, *Vernunftlogik* stands for Hegel in a critical relation with respect to intellectual logic, for at least two reasons. First because it is wider, insofar as it is more complete; second because it is engaged in a critical reflexion on intellectual logic, at the point of revealing that the latter betrays its commitment to truth – in other words: *forms do not always preserve truth.*

---

[40] For Goble (ed.) 2001, 1 "logic supports philosophy and philosophy feeds logic. They join, the result is philosophical logic". Similarly, in 2002, 2f. Jaquette underlines the reciprocity between logic and philosophy: "the role of logic in philosophy has been both tool for the precise expression of arguments and source of philosophical puzzles and paradoxes […] there is reciprocation between logic and philosophy […] in that conceptual housekeeping in the philosophy of logic about […] the semantics of truth, existence […] has contributed directly to refinements in the foundations and superstructure of symbolic logic".

[41] This aspect is stressed by Burgess 2009, vii ff. He writes that philosophical logic is to be understood in non-classical terms, as an extension or alternative to classical logic: "philosophical logic […] is just the part of logic dealing with proposed extensions of or alternatives to classical logic" (vii). Burgess writes that "philosophy of logic […] is no more to be confused with philosophical logic than is history of geology with historical geology. Philosophical logic is a branch of logic, a technical subject" (viii).

[42] See the essays collected in Marconi (ed.) 1979a, as well as Günther 1978, Priest 1989, 388–415, d'Agostini 2000 and Id. 2009, 203–223, Berto 2006, Ficara 2013, 35–52, Bordignon 2014, Ficara 2014a, 29–38.

These are the motivating considerations of philosophical logic as both critical analysis of logic, and creation of *extended* or *alternative* non-classical systems.

Philosophical logic as non-mathematical logic – In all accounts there is a certain tendency to think that philosophical logic is distinct from mathematical logic or from logic generally speaking. The idea of a certain divergence between mathematical and philosophical logic has been stressed by continental philosophers (with reference to the transcendental-idealistic account of logic, or to Husserl's phenomenological conception of logic), but is not ignored in the analytical tradition. In the analytical account logic is mathematical logic in at least two senses: because it consists of creating mathematical languages or models[43], and because it is specifically interested in mathematical language. Now philosophical logic could be distinct from mathematical logic in the second sense: because it is mainly interested in natural language. The basic (Russellian) conception of philosophical logic as an attempt to grasp and express natural forms is thus conserved, in this account, except that eminent "forms" were, for Russell, those given by mathematical logic. Evidently, also in this case, Hegel's conception is incompatible with the Russellian one, exactly for the reason indicated above: because there was no mathematical logic, in Hegel's times.[44]

It should be noted that different interpretations of the idea of philosophical logic (and more generally of the very same idea of "logic" – see Shapiro 2015) are mainly related to the last part of Russell's quotation, that is: "to extract this knowledge from its concrete integuments, and render it explicit and pure". What this means is controversial. Should we produce mathematical models, in order to "extract" the forms of validity? Does "formalisation" really mean what Russell calls "extracting forms", or rather "to construct" or "to create" an ideal language, with a semantic model, an interpretation, etc., in a word: a *logical system?* In doing this, should we use classical methods, principles and conventions, such as truth-functionality, non-contradiction, excluded middle?

In this sense, Hegel's ideas about the relations between natural logic and logic as theory, in their strict affinity to Russell's preliminary account, may help us to come to terms with the "clash" between different conceptions of logic, philosophical logic, and possibly also of philosophy, so that it is even possible to gather most (if not all) hypotheses into a more general view.

---

43 See on this the account of "philosophical logic" provided by Horsten/Pettigrew (eds.) 2014.
44 On the incompatibility between Hegel's logic and mathematical logic see Peckhaus 1997, 122. On Hegel's critique of the adoption of the mathematical method in philosophy see Apostel 1979. On Hegel's appraisal of the mathematical account of the infinite see here 4.3., as well as Moretto 1984 and Ficara 2014b, 59–65.

3.3 Conceptuality and the philosophical approach to logic

It has been often stressed that one reason for the incommensurability between Hegel's logic and contemporary (modern) conceptions is that Hegel's logic – like any other logic in the tradition – was "conceptual", while from Frege onwards logic is "propositional".[45] Surely, there are some differences in this respect. But what is interesting to note now is that, as soon as we focus our attention on the notion of philosophical logic (in the Russellian-Hegelian sense), these differences seem to be less relevant.

What is worth pointing out is that the idea of "conceptuality" as a definite requisite of the philosophical approach to logic is deeply engrained in the analytical development of logic.[46] Russell's first conception of philosophical logic (extracting forms) naturally calls forth the notion of *conceptual analysis*. What is more, Russell's very notion of analysis turns out to be what gathers all the previous conceptions of philosophical logic.

Conceptual analysis, as developed in Russell's theory of descriptions (1905) and in his conception of logical atomism, is a philosophical "extension" of formal logic, whose natural necessity is given by the need for "criticising and clarifying notions which are apt to be regarded as fundamental and accepted uncritically" (Russell 2010, 147). Russell's idea has been variously developed in the analytical tradition. His intuition on the connection between ontological problems and quantification has found in Quine's meta-ontology a very influential expansion and explication.[47] However, and at the same time, many authors have criticised it. From the late Wittgenstein onwards the connection between "logic" strictly speaking and philosophy of language has been questioned and rejected.[48] But what counts to note now is that some "natural" move from logic to philosophical logic, and to conceptual analysis, is to be seen also in the same roots of the most formal account of logic. One would say that in Frege's and Russell's perspective on logic there is something that calls for *conceptual*

---

[45] See on this among others Tugendhat (1970, 152) who writes "Hegel shared the prejudice of the logic of his times according to which judgements are composed by concepts, and the speculative logic that he developed is a logic of the concepts and determinations and systematically violates Frege's view that the primary logical and, one could add, also ontological unity [...] is the sentence". For an analysis of Hegel's view on truth-bearers within contemporary logic see Part III below.
[46] See Frege 1891, now in Frege 2008, 1–22. See also Tugendhat/Wolff 1993, 127 ff.
[47] Quine 1948, see also Chalmers/Manley/Wassermann 2009.
[48] The counter-movement to the Russellian program called *ordinary language philosophy* was very influential from the 1950es until the 1970es.

philosophy, though without renouncing the formal – mathematical – apparatus. That this aspect is outright Hegelian in spirit is evident if one considers Hegel's own ideas about "philosophical logic".

In Hegel's view, as we have seen, philosophical logic itself is the operation (identical to Russell's idea), of extracting the form (the concept) from common language and reasoning (common representations). Significantly, Hegel's account here diverges from the Russellian account in at least one aspect: that for Hegel *"the form" is "the truth"* about the matter at stake.

3.4 Concepts and the forms of truth

Hegel takes for granted that between "forms" and "truth" there is a strict relation.[49] As previously mentioned, for Hegel truth is what grounds the enterprise of "extracting" and exploring forms. What is specifically Hegelian is the idea that the task of extracting forms (concepts) from common language and reasoning (common representation), makes sense *insofar as they are true*, rather: they are *the truth*, which means in Hegel's (Kantian) terms: that they are the very possibility conditions of achieving the truth about any fact or event.[50] In this sense, the analysis of forms for Hegel is the path in order to find the effective truth of what we are thinking, believing or knowing in the case under attention.

It is important to note that the aim of "criticising and clarifying notions" postulated by Russell has no justification except than the idea of increasing the domain of our knowledge *about concepts*, so improving our use of them. But Russell does not openly say *why this ought to be performed by the creation or the discovery of a formal language, rather than by simply using the clarifying resources of natural language*. This missed explanation, on Russell's part, is what makes the "ordinary language" philosophers' objection ultimately reasonable.

If we cling to Hegel's account instead, the reason is clear: the justification of the passage from natural language to a (more or less) formal language is perfect-

---

[49] We will see that between the two concepts there is not proper "identity" for him (see here II and III).

[50] The idea of truth in the transcendental and idealistic tradition (as transcendental truth) is commonly (at least starting from Heidegger) held to be incompatible with the sentential and propositional conception. Here in III I show that the thesis of the incompatibility between transcendental and propositional truth is wrong. For Tugendhat this view is even dangerous: Heidegger's idea of transcendental truth as condition of propositional truth, but in itself non-propositional, is a way of "sanctioning arbitrariness" (see Tugendhat 1970, 334 f.).

ly justified. We simply need this passage, not only in order to "clarify" notions, but also in order *to get the truth about reality and our shared life*. This is simply due to the fact that the forms of *das Logische* are what makes us believe or disbelieve, know or ignore, and so what makes us believe in the truth or falsity about things. Not only that, forms reveal the essence of things, their internal structure, and so they allow us to formulate true ideas about them.

Evidently, Hegel endorses here a specific view about forms, which I will analyse in the next part. What I have tried to sketch here is how the connection between "form" and "truth" is conceived in the perspective of conceptual analysis, and how this connection may justify the enterprise of "logic" as we can still intend it.

Many things remain to be said about Hegel's conception of philosophical logic (and of logic generally intended) with respect to contemporary logic. What is worth noting now is that Hegel's account can dialogue perfectly with our views about logic, and about the task of philosophy with respect to logic. Not only that. The dialogue may be profitable for us. More specifically, the method of philosophical logic in Hegel's view is guided by truth in a peculiar way, a way that I address in the following chapters, and that is somehow underrated in current accounts. For now, one point can be stressed that has emerged so far: while the task of philosophical logic for Russell is "extracting the forms of thought from their concrete integuments", for Hegel the task is "to extract the *forms of true thought* from their concrete integuments", and to show how these forms are the *norm of truth* of the "concrete representations" of things.

Summary

The first part is devoted to Hegel's general idea of logic and its meaning within contemporary philosophical logic. My aim is to dispel some common prejudices, such as "Hegel's logic is identical to metaphysics, hence it has nothing to do with logic commonly intended (as theory of valid inference), which is ontologically neutral".

In **Chapter 1** I consider some Hegelian concepts that are revelatory with respect to Hegel's own view on logic, and have no (or non literal) equivalent in English: *das Logische* – the "logical" (different from *die Logik* – logic), *Verstandeslogik* (intellectual logic), and *Vernunftlogik* (rational logic). The analysis of these concepts, which mark cornerstones of Hegel's general idea of logic, shows that Hegel's view is perfectly reasonable, and not "curious", idiosyncratic or unrelated to our common view of logic, as many authors suggest.

Hegel is probably the first author to use the expression *das Logische*: it refers to the field of logical forms as *facts deposited in our natural language and reasoning*. For Hegel we use forms in our life, actions, decisions, discussions and interactions, without even knowing it, and without awareness about them. *Die Logik* is logic as discipline, which identifies, isolates and fixes the forms that are "unconsciously busy", making them the object of our inquiry. Hegel suggests that there can be a clash between natural logic and logic as theory: sometimes natural logic goes against the rules fixed in the manuals. Interestingly, the case he is most concerned about is the opposite one: when logic as theory clashes against our natural order of thought, when logic as discipline prescribes as valid arguments that are simply wrong, i.e. do not convey truth. Here the distinction between *Verstandeslogik* and *Vernunftlogik* comes into play: *Verstandeslogik* is the logic that fixes forms of valid inference without questioning their validity; *Vernunftlogik* (i.e. dialectical or speculative logic) questions the validity of the rules, when they happen to violate our good, natural way of thinking.

In **Chapter 2** I examine the famous thesis, at the centre of many controversies on the possibility of integrating Hegel into the canon of the history of logic and philosophical logic, "logic coincides with metaphysics". On the basis of a detailed analysis of Hegel's texts, this idea assumes a precise meaning, which can be articulated into three theses: 1) the *Kantian thesis:* the *structures/principles/forms of being*, which constitute the research field of pre-Kantian ontology, are (according to the transcendental turn, which Hegel shares) *structures of our thought about reality*, and are to be dealt with in the context of the discipline we call logic (the theory about the most general forms of thought). 2) the thesis about the *objectivity of thought:* the forms of thought (the propositional, conceptual and inferential forms) are expression of reality, of how things stand. If I identify *modus ponens* as form of valid inference I admit that things/reality is structured and can be structured according to *modus ponens*. 3) the third thesis concerns the *interplay between logic and metaphysics:* logic and metaphysics are inseparable at the natural level, and our natural, everyday thought about reality (and the forms of reality) is intersected with our thought about thought (and the forms of thought) and undistinguishable from it (Hegel writes "natural logic is natural metaphysics"); the two realms of forms are separated when they become research fields of different disciplines, but their natural connection is restored or kept within Hegel's *Vernunftlogik*. On this basis it is possible to address the question: "what kind of logic is Hegel's logic?" from a contemporary perspective. Clearly, Hegel's logic is not a mere list of forms of valid reasoning, but it is a philosophical, that is, in Hegel's view: critical and complete analysis of the forms, and a consideration of the relation between forms of thought and forms of reality. In this sense, Hegel himself also calls his dialectical logic "philosophical logic", coining this very expression. Hence this expression is the common basis in order to assess the actuality of Hegel's position.

In **Chapter 3** I present some leading contemporary definitions of "philosophical logic", highlighting their derivation from Russell's conception of philosophical logic in 1914. For Russell, in philosophical logic we "extract the forms" deposited in our everyday language and reasoning "from their concrete integuments". Russell's program is surprisingly close to Hegel's view. For both thinkers philosophical logic is linked to a view of forms as facts deposited in human life and thought, and coincides with the "extraction" of such facts from their concrete integuments. For both thinkers philosophical logic has a (more or less explicit) connection with conceptual analysis. There are, however, two main divergences between the two ideas: while for Russell (and the philosophers who appeal to Russell) there is a link between mathematical and philosophical logic (the forms are those expressed in the formal language of the *Principia Matematica*), for Hegel the unique logic is the philosophical logic, and the language expressing the forms is the same natural language, and not the symbolised mathematical one. The second divergence concerns the link Hegel establishes between philosophical logic and truth, and his idea of forms as *forms of truth*. For Hegel forms are *possibility conditions of true thought*, and philosophical logic has the task to inquiry into the truth of what we say, do and think. In contrast, the aim of philosophical logic for Russell (and the philosophers inspired by him) is to make our ideas clear, and to solve philosophical problems.

II Form

Logical thoughts are not an *only* [ein *Nur*] against all content, but every other content is only an *only* [nur ein *Nur*] against them [...] [logical thoughts] are the [...] ground of everything. (Hegel Werke 8, 85/Hegel 1991, 59)

The form of an inference, as also its content, may be absolutely correct, and yet the conclusion arrived at may have no truth, because this form as such has no truth of its own. But from this point of view these forms have never been considered. (Hegel Werke 19, 240/Hegel 1894, 222)

In the preceding part of the book some aspects of the Hegelian conception of "logical forms" have emerged, although the focus was on the reconstruction of Hegel's general conception of logic. Here I want to explore Hegel's ideas about the concept of "form" in more detail. The last chapter closed hinting at it, by stressing that for Hegel the task of philosophical logic is, in Russell's words, "to extract the forms of thought from their concrete integuments" in natural language and thought (natural logic). However, we also saw that the meaning of Hegel's philosophical logic is not exhausted by the task of "making natural language and thought clear". This happens also because Hegel endorses a specific concept of form, and consequently of the adjective "formal" in the expression "formal logic". Its clarification is the explicit aim of this chapter. As I will show, Hegel's conception of "form" and "formal" entails a critique of the formalistic views on logic typical for the philosophies of logic of his times, according to which forms are "only forms". In this perspective, Hegel's conception of forms and formal logic corresponds to what I call an operation of "empowering forms".[1] In this perspective, Hegel's talk about the "absolute form" can be clarified.[2]

---

[1] See also Ficara 2019, 15–26.
[2] What it means that Hegel's logic is science of the "absolute form" is clarified by Nuzzo 1997, 50 ff. See also De Vos 1983 and Id. 2006, 210. Nuzzo 1997, 50 ff. claims that Hegel rejects the idea of logic as abstract and formal discipline. If "the logical" is conceived as form separated from every content, then forms are deprived of truth and formal logic is reduced to a logic of falsity, and has no scientific value. Hegel, in contrast, conceives forms dialectically, showing that forms are always forms of something, and contents are always, as such, formed. In this perspective we can understand Hegel's talk about the absolute form, and of logic as science of the absolute form. Nuzzo 1997, 52 writes that thought as form is, for logic, an "absolute" dimension insofar as it is not grounded on something else external to thought. In it, thought begins from itself, proceeds in itself and develops in immanent and autonomous way the complete system of its own manifold determinations. In this sense, Hegel's claim that the absolute form is the absolute truth is to be linked, as Nuzzo shows, to Hegel's idea of the auto-production and auto-determination of thought ("pure" and "absolute" means: the field of logic is only thought, detached from everything else). I share Nuzzo's view, but in the following pages I stress an aspect that, up to now, has been overlooked by the interpreters. In my view, stressing this aspect is important in order to prevent possible misinterpretations of Hegel's position as panlogism, or as metaphysical idealism, i.e. the idea that there is no external reality, and that reality is produced by thought (for a clarifying explanation of Hegel's notion of empirical or external reality see Nuzzo 2003, 171–187). In the following pages, in particular in my analysis of Hegel's reading of what I call Kant's "formalistic argument", I show that what Hegel criticises is not the formal character of logic or the discipline "formal logic", but rather the formalistic philosophies of logic of his times, i.e. the views according to which "truth concerns only the content, and formal logic abstracts from every content, hence formal logic has nothing to do with truth (it cannot provide a material criterion of truth)". I show that what Hegel argues against is the thesis "truth con-

Many authors stress that Hegel's attitude toward formal or common logic is irretrievably critical.[3] Others highlight that Hegel's standpoint is "ambivalent" (Krohn 1972, 57) since Hegel criticised formal (intellectual) logic, but also considered it as a fundamental endeavour. What I have tried to show, in contrast, and will further highlight in the following pages, is that the critical views that are usually traced back to Hegel are not Hegel's own views, but rather common theses Hegel recalls in order to present the diffused scorn of logic typical of the philosophy of his times, a scorn that Hegel himself does not share at all. Hegel defines it "barbaric" (Hegel Werke 6, 375/Hegel 1969, 682).[4] Formal, intellectual logic is not per se despicable, for Hegel. Its content are the forms of truth, forms that are at the very basis of our life, thought and action:

> [T]he several forms of syllogism constantly exert influence on our knowledge. If any one, when awaking on a winter morning, hears the creaking of the carriages on the street, and is thus led to conclude that it has frozen hard in the night, he has gone through a syllogistic operation – an operation which is every day repeated under the greatest variety of complications. (Hegel Werke 8, 335/Hegel 1991, 260)

Hegel even recalls that being aware about the forms we always use is important, for many reasons, first of all for pedagogic reasons, i. e. for educating human beings to the evaluation of arguments, and to critical thought. In the *Subjective Logic* he writes:

> But without going into this aspect of the matter which concerns the education [...] and, strictly speaking, pedagogics, it must be admitted that the study of the modes and laws of reason must in its own self be of the greatest interest – of an interest at least not inferior to an acquaintance with the laws of nature. (Hegel Werke 6, 374 f./Hegel 1969, 682)

---

cerns only the content". Hegel recalls that truth (in its basic meaning as correspondence, a meaning at the very basis of the logical enterprise) concerns the *link* between form and content, and, since logic deals with forms, it is, in principle, perfectly capable of giving an account of truth. Insofar as it does not give the meaning of the forms for granted, and adopts a critical stance towards them, logic is able to make forms apt to express truth. In this way, I show that the interpretations of Hegel's logic as metaphysical idealism, or panlogism, i.e. the view that there is no external reality, is wrong. Hegel's position is better grasped in terms of speculative empiricism, i.e. as the view according to which external reality, as soon as it becomes the object of philosophy and logic, is an analysed reality, a logically structured, rational reality.

[3] See Ritter/Gabriel/Gründer (eds.) 1971 ff., vol. 5, 358. See also Peckhaus 1997, 120 ff.
[4] On Hegel's critique of the romantics' critique of intellectual, syllogistic logic see Krohn 1972, 56.

What Hegel sharply criticises is, in contrast, the way in which the subject of formal logic (the syllogistic forms) is dealt with in the handbooks of his times:

> [T]he most merited and most important aspect of the disfavour into which syllogistic doctrine has fallen is that this doctrine is a *concept-less* occupation with a subject matter whose sole content is the *concept* itself. (Hegel Werke 6, 377/Hegel 1969, 684)

It is an arid, "concept-less" treatment – the syllogistic forms *are* the conceptual realm and are presented without any trace of conceptual thought. For this reason, Hegel states that the forms are reduced to an "ossified material", and logic is a "ruined building". In what follows, I examine these Hegelian views in more detail.[5]

The meaning of "form" and "formal" is the subject matter of many debates both in the history of logic and in contemporary philosophical logic.[6] In what follows, I will show how in Hegel's texts themes pertaining to standard discussions on logical forms (the idea of "forms" as both structures and rules,[7] the legitimacy of the attempt at expressing logical forms using symbols[8]) coexist with elements that are peculiarly Hegelian (the idea that we need to "empower" forms, that logical forms "live" and interact with each other). In this respect, we will see that Hegel's conception of forms is Aristotelian: forms express the essence of things, their concept, the universal, whereby universals do not constitute a separate realm, but are immanent to reality. They are the active, living principles of things. At the same time, Hegel shares the Kantian account according to which forms are the product of both abstraction and semantic ascent.[9]

---

[5] See for a detailed reconstruction of Hegel's standpoint on syllogistic forms Schick 2003, 85–100.

[6] For a clear overview see Dutilh Novaes 2011, 303–332. See also Mac Farlane 2000, Peckhaus 1997, Gabriel 2008, 115–131.

[7] Dutilh Novaes 2011, 306 distinguishes between two basic meanings of the adjective "formal" in the expression "formal logic": formal as pertaining to forms and formal as pertaining to rules, whereby she individuates 5 declinations of the first meaning, all based on the insight that: "What is form, and formal, is what remains once matter is removed (abstracted from)" and that there are 5 basic meanings of "matter" (as thing, subject matter, meaning, content, subclass of the terms of an argument). For the definition of "logical form" of a statement or sequence of statements as underlying structure of the statement (or sequence) see Cook 2009, 177.

[8] Cook (2009, 177) reduces the different meanings of "formal" to two: "formal" as "pertaining to the structure of sentences or arguments" and "symbolic". Read (1995, 61) underlines that the use of "formal" as synonym of "symbolic", though common, is inappropriate.

[9] For an overview on the meaning of "form" and "logical form" see Mittelstraß 1980 ff., 657 ff. For an analysis of the link between Hegel and "common" (or formal) logic see Krohn 1972, Hanna 1986, 305–338, Nuzzo 1997, Nuzzo 2014, 257–273, Redding 2014, 281–301, Gerhard

In what follows I first (in Chapter 4.) examine Hegel's assessment of formal logic from a historical point of view, focusing on the conception of logical form emerging from Hegel's interpretation of Aristotle, Leibniz and Kant. Then (in Chapter 5.) I consider Hegel's statements on formal logic in the *Science of Logic* and the *Logic* of the *Encyclopaedia*, recapitulating and highlighting the main Hegelian theses on logical forms. I conclude (in Chapter 6.) hinting at the meaning of the Hegelian standpoint within contemporary conceptions of "logical form".

2015, 5 – 12. Hanna 1986, 306 claims that Hegel neither merely criticises common logic nor denies its legitimacy. He rather "preserves the entire edifice of common logic while still using the critique of the latter as a motivation for its own self-development towards a more comprehensive and radically new sense of logic. Many of the misunderstandings of Hegel's logic are based precisely on confusions concerning the equally critical and conservative character of Hegel's treatment of the common logic". Hanna 1986, 307 also remarks that Kant, in contrast, did not see the common logic as ontologically naïve and undeveloped, but rather as a well-grounded, necessary propaedeutic and foundation of his transcendental logic. For Schick 2018, 459 the content of Hegel's (subjective) logic is the same as the subject matter of traditional logic – differently from traditional logic, which, for Hegel, is "empirical" and unscientific, Hegel's logic is scientific.

4 Hegel on the history of formal logic

4.1 Aristotle

Hegel's satirical observations about the dullness of ordinary logic are well known. Famous is the Jena aphorism:

> All men are mortal: Caius is a man; thus he is mortal. I at least have never thought such platitudes. It is said to happen internally, without us being aware of it. Well, much happens internally, urine formation etc. (Hegel Werke 2, 541)

Less familiar are perhaps Hegel's detailed remarks on Aristotelian logic (and derivatively on the formal logic of his time, which was, in its core, Aristotelian) in the *Lectures on the History of Philosophy*.[10] Here Aristotle's work on logic is considered to be of fundamental importance for both philosophy and the subsequent logic.

> Aristotle has rendered a never-ending service in having recognized and determined [this becoming conscious about the activities of the abstract intellect], the forms which thought assumes within us. For what normally interests us is concrete thought, thought immersed in external intuition; these forms are sunk within it, constitute a net of eternal activity; and to fix, to bring to consciousness those fine threads which are drawn throughout everything – the forms – is a master-piece of empiricism, and this becoming conscious about them is absolutely valuable. (Hegel Werke 19, 237/Hegel 1892 ff., vol. 2, 219)[11]

---

10 On Hegel's reading of Aristotle's logic see Mignucci 1995, 29–50. Mignucci 1995, 46 recalls that for Hegel logic is "description of forms of thought" and is in this respect perfectly compatible with the Aristotelian conception. According to Mignucci this definition cannot be taken as sign of Hegel's psychologism due to the fact that Hegel, exactly like Aristotle, uses it to underline that "the syllogisms are controlling structures for the deductions developed by the different sciences".

11 I have partially changed the 1892 translation, which is not close enough to the German original (reported below with the 1892 translation following). In the English translation the focus (which is evident in the German text) on the connection between unconscious natural logic and logic as theory gets lost, since the passages on *becoming conscious (Bewusstwerden) about the forms* and on *bringing them to consciousness (zum Bewusstsein zu bringen)* are omitted. Not only that, "dies Bewusstsein" is translated with "this knowledge".

Es ist ein unsterbliches Verdienst des Aristoteles, dies Bewusstwerden über die Tätigkeiten des abstrakten Verstandes, diese Formen erkannt zu haben, die das Denken in uns nimmt. Denn was uns sonst interessiert, ist das konkrete Denken, das Denken versenkt in äußere Anschauung; jene Formen sind darin versenkt, es ist ein Netz von unendlicher Beweglichkeit; und diesen feinen, sich durch alles hindurchziehenden Faden – jene Formen – zu fixieren,

The passage entails an evident reference to the interplay between *das Logische* (as both the field of forms that are there, in our life, action, thought, and the unconscious activity of thinking, acting, inferring according to those forms) and *die Logik*, i.e. logic as theory, which brings to consciousness, fixes, and determines those forms. Hegel claims that Aristotle managed to fix and determine the forms, and that his accomplishment is a "master-piece of empiricism". It is interesting to note that by "empiricism" Hegel does not mean Humean empiricism, but what he, in other places, calls "Aristotle's speculative empiricism". Hegel thus hints here at the fact that the *empiria*, the field of experience to which Aristotle refers, is a logico-linguistic one, that of the forms of thought.[12]

Aristotle managed to identify, fix and enumerate for the very first time the forms of judgements and inference that, as Hegel contends, are "sunk within thought and language".[13] The logic manuals of Hegel's times take everything from Aristotle and develop new insights only "as far as details are concerned [...] but the truth is to be found with Aristotle" (Hegel Werke 19, 238/Hegel 1892 ff., vol. 2, 220). Even if Hegel recognises that Aristotle's logic may appear to be dry and lacking in content,[14] he also observes that formal logic is a fundamental science, which concerns everyone and should be studied by everyone.

> However little this logic of the finite may be speculative in nature yet we must make ourselves acquainted with it, for it is everywhere [in every science]. There are many sciences,

---

zum Bewusstsein zu bringen, ist ein Meisterstück von Empirie, und dies Bewusstsein ist von absolutem Wert.

Aristotle has rendered a never-ending service in having recognized and determined the forms which thought assumes within us. For what interests us is the concrete thought immersed as it is in externalities; these forms constitute a net of eternal activity sunk within it, and the operation of setting in their places those fine threads which are drawn throughout everything, is a master-piece of empiricism, and this knowledge is absolutely valuable.

12 On Aristotle's speculative empiricism see Hegel Werke 19, 145 ff./Hegel 1892 ff., vol. 2, 131 ff. Hegel famously defends a peculiar interpretation of Aristotle's philosophy, underlining the substantial continuity between Plato's and Aristotle's thought. In particular, he views Aristotle as the philosopher who manages to set the Platonic idea in motion. On Hegel's Aristotelianism see Verra 2007, 349–370. On Hegel and Aristotle see Ferrarin 2001. On the continuity between Hegel's and Aristotle's concept of thinking see de Laurentiis 2002, 263–285 and Id. 2005.

13 Hegel underlines that to the forms individuated by Aristotle also belong the categories, i.e. concepts expressing the fundamental structures of being, which in the philosophy before Hegel (not in Kant) were treated separately, not in handbooks on logic but of metaphysics (ontology). On the Hegelian (and Kantian) view of metaphysics as essentially belonging to logic see the previous chapter.

14 As Hegel writes: "[the syllogistic forms] may not seem to serve their purpose of discovering the truth" (Hegel Werke 19, 238/Hegel 1892 ff., vol. 2, 220).

> subjects of knowledge etc. that know and apply no other forms of thought than these forms of finite thought [...] mathematics, for instance, is a constant series of syllogisms; jurisprudence is the bringing of the particular under the general, the uniting together of both these sides. (Hegel Werke 19, 240/Hegel 1892ff., vol. 2, 222f.)

Similar observations can be found in many other writings. I have mentioned Hegel's conviction that the "logic of the understanding" is the very condition of the philosophical, rational one. But it should also be mentioned that Hegel genuinely appreciated the importance of "ordinary logic", an "empirical" discipline that, as he suggests, is more worthy than other empirical kinds of research:

> It is held to be a worthy endeavour to gain a knowledge of the infinite number of animals, such as one hundred and sixty-seven kinds of cuckoo [...] or to make acquaintance with some miserable new species of a miserable kind of moss [...] or with an insert, vermin, bug etc. in some learned work on entomology [but] it is definitely more important to be acquainted with the manifold kinds of inferences than to know about such creatures. (Hegel Werke 19, 238/Hegel 1892ff., vol. 2, 220)

In fact, Hegel's critique of both ordinary and Aristotelian logic is rather addressed to some shared and superficial *views on logic*, and second and derivatively to some actual problems concerning the Aristotelian conception itself. In both cases the concept of "form" plays a crucial role.

As to the first point (which we will see better considering Hegel's reading of Kant's view on formal logic) Hegel criticises the view, typical for the "philosophy of logic" of his times, according to which the formal character of logic means that inferential *forms are lacking in content*. In this respect, logic in its "ordinary" account diverges from the Aristotelian conception. Hegel stresses that according to Aristotle logical forms express the essence of things. As we have seen, Hegel accepts and reintroduces the Aristotelian and ancient Greek conception, in virtue of which there is a natural connection between *logos* and *on*, and the analysis of forms is to be connected to truth, and to the examination of how things really and truly are:

> Concepts of the understanding or of reason constitute the essence of things, not certainly for that point of view [the current view of logic in Hegel's times], but in truth; and also for Aristotle the concepts of the understanding, namely the categories, constitute the essence of things. (Hegel Werke 19, 240/Hegel 1892ff., vol. 2, 222)[15]

---

[15] Hegel refers here to "concepts of the understanding or of reason" and not primarily to "forms": but "categories" are to be intended as "forms" in Kant's sense: see 2.1.

Since logical forms express the essence of things, they are not things' disembodied or schematic version, but rather their internal principle and reason for being. They are that in virtue of which things are the things that they actually are. Forms do not lack content, as their content *is* the essence of beings. This is precisely the Aristotelian conception of "form".

As to the second point, i.e. the critique of Aristotle's logic itself, Hegel stresses that in Aristotle's theory of syllogism the forms do not manage to meet their claim to be forms of truth (Hegel Werke 19, 235/Hegel 1892ff., vol. 2, 217) and that

> [T]he true cannot be found in these forms. But it must be remarked that Aristotle's logic is not by any means founded on this relationship of the understanding [*Verstand*]; and that Aristotle does not by any means proceed in accordance to these syllogistic forms. (Hegel Werke 19, 241/Hegel 1892ff., vol. 2, 223)[16]

In other words, Aristotle does not use what he himself proposes as a general form of reasoning. Aristotle fixes the complex apparatus of the syllogistic system, but does not apply it in his writings. What is the point in giving such a complex system? Hegel's diagnosis is that the system itself has lost its *raison d'être*, that is, its relation to truth. These forms are not the forms of truth. Formal logic, as it is developed by Aristotle, does not manage to meet the claim, intrinsic to the logical enterprise, of expressing the most general form of truth.

Oddly enough, according to Hegel this failure is ultimately due to the fact that Aristotelian forms are not "formal" enough. Hegel argues that the reason for the deficiency of Aristotle's logic is not, as the common view would have it, that the forms of inferences are "only forms". Rather, the problem is that "*form is lacking to them* [my emphasis], and that they are in too great a degree content" (Hegel Werke 19, 239 (but see also 240)/Hegel 1892ff., vol. 2, 222). As will become clear in dealing with the relation between Kant's and Hegel's views on logic, what Hegel wants to emphasise is that these forms in Aristotle are just enumerated and fixed, without any reflection about their relations to each other, and without any explanation of what they are and that to which they are to be referred. They have lost their use. Thus Hegel also writes:

---

16 The 1892 English translation is "Aristotle's *philosophy* is not by any means founded on this relationship of the understanding", but I have substituted "philosophy" with "logic" since in the German text Hegel writes "Logik" and not "Philosophie". The English translation supports the common, though wrong, and non-Hegelian, view about the separation between philosophy (along with philosophical logic) and formal logic; while the German text suggests that the two enterprises are inseparable.

> The form of an inference, as also its content, may be absolutely correct, and yet the conclusion arrived at may have no truth, because this form as such has no truth of its own, but from this point of view these forms have never been considered. (Hegel Werke 19, 240/Hegel 1892ff., vol. 2, 222)[17]

So one could say that the way to meet the claim naturally raised by logic to set up universal forms of truth is introducing self-reflection, and self-criticism within *Verstandeslogik* (the Aristotelian logic of Hegel's times), self-reflection and self-criticism that are lacking in both Aristotle's and the view on logic typical of Hegel's times.

Finally what is wrong in the "ordinary" conception of form as well as in the Aristotelian one, is that forms are *isolated*, which for Hegel means: abstract. They are abstracted from their use as "forms of truth", isolated from their nature as "essences of things", separated from each other and conceived in total ignorance of their mutual connection. As Hegel writes, the forms of judgements, syllogisms and concepts are, in ordinary logic, "not considered in their unity but only in their isolation" (Hegel Werke 19, 240/Hegel 1892ff., vol. 2, 222). This also implies that Hegel reacts against the usual way of dealing with the subject matter of logic, according to which the theories of judgements, concepts and inferences are three different logical disciplines, to be treated separately. One central question that, in Hegel's view, needs to be answered is the one about the link between concepts, judgements and syllogisms. This question was not considered by the formal logic of Hegel's time and constitutes one main motivation of dialectical logic. Accordingly, Hegel conceives his dialectical logic as a *unified* theory about the connections between judgements, concepts and inferences.[18]

---

17 I have changed the translation, which is misleading. The German text is "Die Form eines Schlusses, so wie sein Inhalt, kann ganz richtig sein und doch sein Schlusssatz ohne Wahrheit, weil diese Form als solche für sich keine Wahrheit hat" and the English translation "The form of a conclusion, as also its content, may be quite correct, and yet the conclusion arrived at may be untrue, because this form as such has no truth of its own". I have substituted "the form of a conclusion" with "the form of an inference" because by *Schluss* Hegel means here, evidently, "argument" or "inference", and not the conclusion of an argument. For the conclusion of the argument he uses *Schlusssatz*.

18 This has been stressed by Käufer 2005, 259–280, who considers Hegel's view as a fundamental step towards Frege's revolution in logic.

4.2 Stoic Logic

Hegel underlines the difference between the Stoic and the Aristotelian concept of formal logic as follows:

> Logic is to [the Stoics] logic in the sense that it expresses the activity of the understanding as of conscious understanding; it is no longer as with Aristotle, at least in regard to the categories, undecided as to whether the forms of the understanding are not at the same time the essences of things. Rather the forms of thought are set forth as such for themselves. Therefore the question comes in, for the first time, about the correspondence of thought and object or the question of showing a peculiar content of thought. (Hegel Werke 19, 273 f./Hegel 1892 ff., vol. 2, 255)[19]

According to the Stoics the logical field is the field of the forms of thought, while in Aristotle "it was left undecided if the forms of thought are at the same time the essences of things".

For the Stoics the logical forms are principles or criteria of truth. For Hegel, this is a fundamental insight and corresponds to the very construction of a formal logic: "since they made thought the principle [of truth], they built formal logic" (Hegel Werke 19, 273/Hegel 1892 ff., vol. 2, 254 f.).[20] As Hegel writes:

> The Stoics indeed had a system of immanent determinations of thought, and actually did a great deal in this direction; for Chrysippus specially developed and worked out this logical aspect, and is stated to have been a master in it. But this development took a very formal direction; there are the ordinary well-known forms of inference, five of which are given by Chrysippus, while others give sometimes more and sometimes fewer. One of them is the hypothetical syllogism through remotion, "When it is day it is light, but now it is night and hence it is not light." These logical forms of thought are by the Stoics held to be "the unproved that requires no proof." (Hegel Werke 19, 275/Hegel 1892 ff., vol. 2, 256)

Hegel suggests that even if the Stoic insight according to which the forms of logic are laws of truth is of supreme value, it presents a fundamental problem, a problem which was highlighted by Sextus in his critique of Stoicism. The difficulty is that, once we establish the forms of thought as valid, i.e. as principles of truth, then "everything can be taken up into them" because everything, also false contents, can be structured according to those forms:

---

[19] On the importance of Hegel's reading of stoic logic in the *Lectures on the History of Philosophy* for understanding the compatibility between the meaning of "logic" in Hegel and contemporary logic see Redding 2014, 281–301.
[20] On Hegel's concept of truth, and a consideration of its relation to the Stoic one see Part III below.

> Since all given content may be taken into thought and posited as something thought [...] the taking of it up does not help at all; for its opposite may also be taken up and set forth as something thought. (Hegel Werke 19, 274/Hegel 1892ff., vol. 2, 255)

I will go back to the critique of the formalistic nature of the Stoic conception in the context of the analysis of Hegel's critique of Kant's formalism. What is worth noting is that Hegel finds in the Stoic approach to forms an important ally in introducing his idea of logic as the science of "the forms of truth", where forms are not the abstract schemas of minor-major-middle terms, but rather living entities of thought and being. They are living insofar as they lead us to believe and disbelieve, and insofar as they constitute the germinal principles (essences) of the things of which they are forms.

4.3 Leibniz

Hegel's standpoint on the meaning of "formal" in a second (non-natural) sense, as "formalised" or "symbolic", clearly emerges in his reading of Leibniz's *calculus ratiocinator* and *lingua characteristica*.[21] Hegel strongly criticises the attempts made by Euler, Lambert and Leibniz of using symbols and signs in order to express conceptual relations and determinations. In the *Conceptual Logic* (Begriffslogik) of his *Wissenschaft der Logik* he criticizes the very idea of a *calculus ratiocinator*:

> [*Euler*, Lambert] and others, have attempted to construct a *notation* for this kind of relations between conceptual determinations [what is meant here are the relations of contrariety, contradiction, subcontrariety and coordination between predicates] by lines, figures and the like, the general intention being to *elevate*, or rather in fact *to degrade*, the logical modes of relation to a *calculus*. The utter futility of even attempting a notation is at once apparent when one compares the nature of the sign and what it is supposed to signify. (Hegel Werke 6, 293 f./Hegel 1969, 616)

---

21 On Hegel's critique of Leibniz' *calculus ratiocinator* and *lingua characteristica* see Peckhaus 1997, 120ff. as well as Kirn 1985, 40ff. On Leibniz' distinction between *calculus ratiocinator* and *lingua characteristica* at the origin of the two traditions in logic – the Boolean, linked to the idea of logic as *calculus* (Boole 1847) and the Fregean, linked to the idea of logic as language – see Peckhaus 2004, 3–14. On the continuity between Hegel's logic (Hegel's concept of the concept) and Leibniz' conception of the monad see Horn 1982, 131–134. Horn retraces this insight to Glockner (ed.) 1960, 6, who states "Hegel merely accomplished what Leibniz had started".

The uselessness of such an endeavour emerges, Hegel thinks, if one reflects on the difference between the static and finite nature of the algebraic signs and the *fluidity* of what they are supposed to designate, namely the conceptual content.

> It is characteristic of such objects [spatial and algebraic signs] that, in contrast to conceptual determinations, they are mutually *external,* and have a *fixed* character [...] It is therefore quite inappropriate for the purpose of grasping such an inner totality, to seek to apply numerical and spatial relationships in which all determinations fall asunder; on the contrary, they are the last and worst medium which could be employed. (Hegel Werke 6, 294 f./Hegel 1969, 618)

Thus:

> Since man has in language a means of designation peculiar to reason, it is an idle fancy to search for a less perfect mode of representation to plague oneself with [...] It is futile to seek to fix [the concept] by spatial figures and algebraic signs for the purpose of the *outer eye* and an *uncomprehending, mechanical mode of treatment* such as a *calculus*. (Hegel Werke 6, 295 f./Hegel 1969, 618)

Similarly, in the "Philosophy of Spirit" of the *Encyclopaedia*, Hegel stresses the difficulty in Leibniz's project of developing a universal language in a "hieroglyphic" manner. The main problem is that the symbols and signs are not able to express the changes that are typical of thought's development.

> Leibniz's practical mind misled him to exaggerate the advantages which a *completed* written language, formed on the hieroglyphic method would have as a universal language for the intercourse of nations and especially of scholars [...] comprehensive hieroglyphic language for ever completed is impracticable. Sensible objects no doubt admit of permanent signs, but [...] the progress of thought and the continual development of logic lead to changes in the views of their internal relations and thus also of their nature. (Hegel Werke 10, 273)

Interestingly, Hegel emphasises the difficulties of fixing a *complete* hieroglyphic language, and the fact that the conceptual and logical relations require signs that are susceptible of a continuous revision. In this passage, Hegel underlines the advantages of natural language, which seems to offer a better guarantee of both flexibility and precision.

However, Hegel also seems to acknowledge that *every* language is inadequate to express dialectical conceptual relations, natural language included: "the sentential form is not suited to express speculative truths" (Hegel Werke 5, 93/Hegel 1969, 90). This apparent self-contradiction is promptly dispelled as soon as one considers the nature of the "algebraic" interpretation of language. Hegel says that algebraic signs do not capture the fluidity and multiplicity of conceptual contents. Yet, similarly, one might wonder whether the uniform and fixed "name" we use to denote concepts is really able to capture this variety and movement. This is the reason why forms, in Hegel's view, are not rejected as such, but only in their non-dialectical use. We obtain the true nature of logical forms (their ability to lead us to truth) only if we maintain the dialectical (critical, or rational) consideration of them. Such a consideration is what gives back to forms their "living" nature as expressions of the life of thought.

This is why Hegel uses a sort of pre-formalized language in order to express dialectical contents, and also attempts formalizing Fichte's dialectics (in the *Differenzschrift*, Hegel Werke 2, 37 ff.).[22] Not only that, in the famous note on the mathematical infinite in the second section of the *Seinslogik* (Hegel Werke 5, 279 ff./Hegel 1969, 240 ff.) Hegel considers notation in higher analysis as more apt to express philosophical, i.e. infinite and incommensurable, contents than natural language itself.[23]

> The character of the mathematical infinite and the way it is used in higher analysis corresponds to the concept of the genuine infinite. (Hegel Werke 5, 284/Hegel 1969, 244)

For example, if we take a fractional number,

> Such fraction, 2/7 for example, is not a quantum like 1, 2, 3, etc.; although it is an ordinary finite number it is not an immediate one like the whole numbers but, as a fraction, is directly determined by two other numbers which are related to each other [...] The fraction 2/7 can be expressed as 0.285714...etc. As so expressed it is an infinite series; the fraction itself is called the sum, or *finite expression* of it. (Hegel Werke 5, 285 f./Hegel 1969, 244 f.)

In other words, the processual version of the infinite expressed by the fraction 2/7 is 0.285714... *ad infinitum*, the infinite series. However, while the series: 0.285714... is an expression of *bad infinity*, 2/7 represents a *good infinity*. In the fraction the infinite is, as Hegel writes, "all contained" as such; the mathematical symbol immediately conveys the entire development of the infinite series. In contrast, in the progress of the series, which for Hegel is also the traditional philosophical way of expressing the infinite, the infinite is incomplete, it always requires the iterated intervention of an operation. In this respect, Hegel underlines that the use of the fraction is not just a way like any other of expressing a content, but has the specific advantage of expressing it in a more precise manner. Here we can clearly grasp the notion of "living form" that is typically Hege-

---

[22] See also Hegel's enthusiasm about mathematical notation in higher analysis, Hegel Werke 5, 279 ff./Hegel 1969, 240 ff.
[23] See also Moretto 1984.

lian.[24] The fixity of the sign '2/7' contains and expresses the infinite movement of 0.285714...

Hegel's observations on the attempts (carried out by Leibniz, Euler and Lambert) to express thought using figures and signs would, if taken univocally, ultimately condemn every instrument of communication, natural language included. And yet, Hegel explicitly claims that natural language and the sentential form, though limited, is all we have in order to express dialectical contents. Moreover, Hegel's analysis of the mathematical infinite even shows that he actually held mathematics, and higher analysis in particular, as better apt to express dialectical conceptual contents.

This suggests that Hegel's critique of the use of symbols should not be seen as an interdiction but rather as a cautionary indication about *how to use both natural and symbolic language in order to express dialectical contents.*

4.4 Kant

In the *Introduction* to the *Encyclopaedia* and in the *Vorbegriff* to the *Logic* of the *Encyclopaedia* Hegel refers to the forms isolated by formal logic and Kantian transcendental philosophy in terms of "universality and necessity". In particular, Kant famously defines in terms of "universality and necessity" the field of *a priori* structures that constitutes the subject matter of transcendental logic. Interestingly, the German word for "universality" is *Allgemeinheit*, which can be translated with both "generality" and "universality". "General/universal" is used by Kant also interchangeably with "formal" in the expression "formal/general logic". Notably, the formulation itself "formal logic" also dates back to

---

24 The conception of "living form" emerging from Hegel's discussion of formal logic is also typical of Goethe's writings. Goethe criticised scholastic logic complaining about its violent treatment of pure thought, "tearing everything apart", "classifying and reducing everything". In it the organic nature of pure thought gets lost (see Goethe 1808, vers. 1944f and 1936–1940). Interestingly, Gabriel 2008, 121ff. underlines that an analogous organic view of forms was at the basis of Frege's *Begriffsschrift*. Gabriel traces Frege's account back to Trendelenburg's interpretation of Leibniz, and stresses that Trendelenburg was, as many in his times, influenced by Goethe. Cassirer develops his idea of "symbolic forms" accordingly (hinting at Hegel's and Schelling's conception of "symbol" in their works on aesthetics, and with explicit reference to Goethe). According to Cassirer's idea, forms, as what "stands amid us and the real objects", do not only refer to "the distance between us and the world, but they constitute the only possible, adequate mediation and the medium through which every [...] being is graspable and comprehensible" Cassirer 1982, 132f.

Kant.[25] In Kant's conception, in particular, the link between universality, necessity and formality is perfectly clear, and the double *descriptive* and *normative* nature of logical analysis comes into view. In the *Jäsche Logic* we read:

> Everything in nature [...] takes place according to rules, although we do not always know these rules. Water falls according to laws of gravity [...] the fish in the water, the bird in the air, moves according to rules. All nature, indeed, is nothing but a combination of phenomena which follow rules; and nowhere is there any irregularity. (A 1)

Everything (natural events such as the bird's flying and the fish's swimming) happens according to laws, even if we are not always aware of it. Clearly, Kant endorses the same Hegelian-Russellian view according to which the forms (Kant refers to *rules*) are sunk in reality, and that logic as a discipline arises when we make them the object of our thought.

> If, however, we [...] reflect simply on the exercise of thought in general, then we discover those rules which are absolutely necessary, independently of any particular objects of thought, because without them we cannot think at all. (A 3f.)

Kant thus links the logical enterprise (as the discovery of necessary rules) to thought's reflection on its own modalities. In the *Critique of Pure Reason* Kant connects formality and generality:

> General [*allgemeine*] logic abstracts from every content of knowledge and from the differences between objects and only has to do with the mere form of thought. (B 26)

Hence two aspects: thought's reflection on its own modalities and abstraction from particular contents, are, for Kant, conjoined in logical inquiry. Moreover, insofar as logic considers the mere structure or form, and not the content, of

---

25 See the entry on logic in Ritter/Gründer/Gabriel (eds.) (1971ff.), vol. 5, 375–378. According to Scholz 1959, 14, Kant was the first who called logic "formal". Krohn 1972, 108, highlights that Aristotle, differently from Kant, did not apply the determinations of "matter" and "form" to determine the nature of his inquiries in the *Organon*. The different implications entailed in the two expressions "formal" and "general" are discussed and deepened in the phenomenological tradition. Tugendhat 1970, 39 explains the difference as follows: while "general" refers to the logical field as the result of an abstraction (logic abstracts from the specific contents thought in each case, and deals with the most general forms of thought), "formal" entails a reference to reflection and semantic ascent, i.e. to the fact that we gain access to the logical field not only by *abstracting away from specific contents* but also by *reflecting on the specific modalities* in which we speak and reason. In Kant's and Hegel's use and understanding of "formal" logic both aspects are included.

thought, it discovers those forms that are absolutely necessary because "without them we cannot think at all". Thus abstracting away from every content allows us to find the forms of everything that can be thought and said, of every content. The forms are general (*allgemein*) in the sense of universal, i.e. common to everything that can be thought and said. Hence the formal logical approach is, on one hand, very humble, since it implies the renunciation of asking about the truth of what is thought in each time. On the other, it is very ambitious, insofar as it entails the universal claim to be an inventory of the forms of *every* truth.[26] As such they are necessary since "we cannot think at all without them".[27]

In the *Vorbegriff* of the *Encyclopaedia Logic* Hegel uses Kant's logical terminology, and the formulation "necessity and universality", in the context of a comparison between Hume's empiricism and Kant's critical philosophy.

> [If we consider Empiricism, we see that] in what we call Experience, as distinct from mere single perception of single facts, there are two elements. The one is the matter, infinite in its multiplicity, and as it stands a mere set of singulars: *the other is the form, the characteristics of universality and necessity* [my emphasis]. Mere experience no doubt offers many, perhaps innumerable, cases of similar perceptions: but, after all, no multitude, however great, can be the same thing as universality. Similarly, mere experience affords perceptions of changes succeeding each other and of objects in juxtaposition; but it presents no necessary connection. If perception, therefore, is to maintain its claim to be the sole basis of what men hold for truth, universality and necessity appear something illegitimate: they become an accident of our minds, a mere custom, the content of which might be otherwise constituted than it is. (Hegel Werke 8, 111f./Hegel 1991, 80)

Empiricism recognises the difference between matter and form, but refuses to attribute to forms the qualities of universality and necessity. It holds to perception as the sole source of truth. In this sense, empiricism does not lead to logic or to the recognition of forms as the horizon of universality and necessity, and as the source of truth. It considers forms as a mere accident, or habit.

Critical philosophy, not unlike Humean empiricism, considers experience as the starting point of thought and distinguishes between matter and forms. However, it also considers forms as the source of thought's universality and necessity:

> [Critical philosophy starts from] the distinction between the elements found in the analysis of experience – the *sensible material* and its *universal relations* [...] the reflection mentioned in the preceding paragraph that only *what is singular* and only *what happens* are contained in perception [taken] on its own account [is in critical philocophy combined with the idea that]

---

26 Litt 1961, 261 underlines a similar point.
27 The link between logic's formality and necessity is underlined today in a similar way. Restall 2006, 215 for instance observes: "The necessity of logic is a matter of its *generality*".

> *universality* and *necessity* [...] are found to be present in what is called experience. And because this element does not stem from the empirical as such, it belongs to the spontaneity of *thinking*, or is apriori. The thought determinations or *concepts of the understanding* make up the objectivity of the cognitions of experience. (Hegel Werke 8, 112f./Hegel 1991, 81)

As in empiricism, experience in critical philosophy is the very *beginning* of thought and knowledge. However, critical philosophy finds the character of universality and necessity within experience, it finds universal and necessary thought structures that, though been found within experience, do not *depend* on it, and rather exercise a constitutive constraint on it. Logical forms are of this kind. This means that forms for Kant have a coercive or constitutive impact on experience.

Hegel endorses the Kantian view about the universal and necessary character of logic as both focused on the mere form and able to give an account of everything that can be thought and said, though, as we will see in the following, what Hegel discusses is a *formalistic reading* of this view. So it is possible to state that the universality and necessity thesis holds for both Kant and Hegel, but not for empiricism: logical forms are universal and necessary in the sense that they are the condition of everything that can be thought and said; they are that without which nothing can be said and thought.

Both Kant and Hegel acknowledge that logical forms are necessary in the sense that, as soon as we recognise them, we recognise their binding nature. Insofar as the logical forms *describe* the structure of what we think, abstracting away from the peculiar contents or circumstances, they indicate how we *ought* to think, always and in all circumstances. Insofar as I recognise how I think, how thought works, I also know when it does not work. I realise for instance that from "every football player is biped" and "Socrates is a football player" I can infer that Socrates is biped while from "every football player is biped" and "Socrates is biped" I cannot infer that Socrates is a football player. Thus, the valid form is not only the result of considering how we factually think when we think well, i.e. the description of our good arguments, but also the prescription of how we ought to think. In this sense Kant writes about forms as *rules*, a use that is confirmed in the normative conception of forms in Hegel.

In general terms, Hegel's view about the limits of traditional (also Aristotelian) logic moves along Kantian lines. Kant suggests that a reflection on the genesis of logical forms and their specific relation to the content of knowledge is needed, and conceives transcendental logic accordingly, as an inquiry into the origin of logical forms commonly intended and their relation to content.[28] The

---

28 See Kant B 81.

same, as we have seen, holds for Hegel. However, it is precisely on this point that the divergence between the two thinkers comes into view: *Hegel did not argue (as Kant did) for the separation of transcendental (dialectical) from formal logic, and of formal logic from effective truth.*

The connection of logic and truth marks the main difference between the Kantian and the Hegelian conception. In the section of the *Subjective Logic* of the *Science of Logic* on the *Concept in General* Hegel recalls Kant's observations about formal logic in the *Introduction to Transcentendal Logic* in the *Critique of Pure Reason*. The question at stake is, openly, the relation of logic and truth:

> When Kant in the *Critique of Pure Reason* in relation to logic comes to discuss the old and famous question: *what is truth?* he first of all *takes for granted* as a triviality the nominal explanation, [according to which truth is] the correspondence of knowledge with its object – a definition [which actually has] great, indeed supreme, value. (Hegel Werke 6, 265f./ Hegel 1969, 593)

The definition of truth in terms of correspondence is not a triviality, but something of "*supreme*" importance. The common view (defended by Kant himself) according to which logic, insofar as it is formal, "abstracts away from content" – while truth "concerns the content", and thus "a sufficient and at the same time general criterion of truth cannot possibly be given" (B 84) – is self-contradictory.

Hegel quotes Kant (B 84):

> "What we require to know" Kant goes on to say, "is a universal and sure criterion of any cognition whatever; it would be such a criterion as would be valid for all cognitions *without distinction of their objects*; but since with such a criterion abstraction would be made from *all content* of the cognition (*relation to its object*) and *truth concerns precisely this content*, it would be quite *impossible* and *absurd* to ask for a mark of the *truth of this content* of cognitions." (Hegel Werke 6, 266/Hegel 1969, 593)

The original Kantian argument is in B 82–84 and is reported by Hegel quite faithfully. I suggest to call it the formalistic argument (FA). Its steps are:

(FA) 1. Truth is the correspondence of knowledge with the object.
  2. What we are looking for [in logic] are universal criteria of the truth of all knowledge.
  3. A universal criterion must be valid for all cognitions without distinction of their objects
  4. With such a criterion abstraction would be made from all content of cognition (relation to its object)
  5. Since truth concerns precisely this content to ask for a mark of the truth of this content of cognitions is impossible and absurd
  6. Sufficient and at the same time general criteria of truth cannot possibly be given.

Hegel comments:

> Here, the usual conception of the formal function of logic is expressed very definitely and the argument adduced has a very convincing air. But first of all it is to be observed that it usually happens with such formal ratiocination that it forgets in its discourse the very point on which it has based its argument and of which it is speaking. It is alleged that it would be absurd to ask for the criterion of the *truth of the content* of cognition; but according to the definition it is not the *content* that constitutes the truth, but the *correspondence* of the content with the concept. (Hegel Werke 6, 266 f./Hegel 1969, 593)

The Kantian (and usual) view about the formality of logic "forgets the very point on which it is based and of which it is speaking". In a successive passage Hegel observes that formalism, "as soon as it starts to explain its own position says the opposite of what it intends". He means, in other words, that FA is self-contradictory. In fact, it starts assuming 1., the very definition of truth as *correspondence*, which implies that truth concerns *both* thought *and* content, and then denies it, in stating 5., i.e. that truth *only* concerns content.

Surely, logic is also for Kant the science of the forms of truth. As he writes: "logic, in so far as it expounds the universal and necessary rules of the understanding, must in these rules give criteria of truth" (B 84). The same holds for Hegel. Kant explains that the notion of truth at the basis of the logical enterprise is correspondence: "The nominal definition of truth, that it is the agreement of knowledge with its object, is assumed as granted" (B 83). The same holds for Hegel too. However, Hegel also underlines the *supreme* and hence not exclusively *nominal* value of the classical correspondence truth. Thus, while for Kant the forms fixed by logic are "*only* forms of truth [my emphasis]" and hence "they are quite correct, but are not by themselves sufficient" (B 84) Hegel has a more ambitious idea of logical form. He does not deny that truth concerns the content thought in each case, but stresses that it does not concern *only* the content, but rather the coincidence of thought and the object or content. This means that for Hegel, as well as for Kant, the logician studies truth by focusing on the forms. Yet for Kant forms are "only forms, and thus not sufficient", while for Hegel saying that the forms of logic are *only* forms of truth is not at all a sign of their powerlessness.

In the *Logic* of the *Encyclopaedia* Hegel stresses that the logical laws are not an "*only* against all content [ein *Nur* gegen allen anderen Inhalt]" but rather that "every other content is only an only against them [aller anderer Inhalt ist nur ein *Nur* gegen dieselben]". Moreover, he claims that "[logical thoughts] are the [...] ground of everything ([die logischen Gedanken] sind der [...] Grund von allem)" (Hegel Werke 8, 85/Hegel 1991, 59). As we have seen, insofar as we individuate the forms of thought we grasp the essence of reality. Evidently, in doing logic one will have to work solely on the forms. However, for Hegel this is not a limit. The problem for him is rather whether the forms we fix are *truly* forms of truth:

> Even here the forms which come up for treatment as well as their further modifications are only, as it were, historically taken up; they are not subjected to criticism to determine whether they are in and for themselves true. Thus, for example, the form of the positive judgement is accepted as something perfectly correct in itself, the question whether such a judgement is true depending solely on the content. Whether this form is in its own self a form of truth [...] is a question that no one thinks of investigating. (Hegel Werke 6, 268/Hegel 1969, 594)

We see here that FA is not only wrong (self-contradictory), but also dangerous: it prevents Kant to ask about the legitimacy of extant logical forms. The question whether the forms are true is not asked because truth is said to "depend solely on the content", whereby "we cannot check the content in each case".

5 Hegel on logical forms

What I have suggested in considering Hegel's approach to the history of logical forms is basically consistent with Hegel's theories about the formal nature of logic in the *Wissenschaft der Logik* and the *Logic* of the *Encyclopaedia*. Now I will sum up the results obtained, by direct reference to Hegel's logical texts, and stressing all the points that might be interesting in light of the "modern" conception of logical forms.

5.1 Five theses on forms

1. In the Preface to the second edition of the *Science of Logic* Hegel writes that the forms produced by Aristotelian and earlier logic "must be regarded as an extremely important source [of the *Science of Logic*], indeed as a necessary condition and as a presupposition to be gratefully acknowledged" (Hegel Werke 5, 19/ Hegel 1969, 31). In this respect, the task of Hegel's *Wissenschaft der Logik* is to "empower" the forms traditionally studied by logic and metaphysics (the realm of thought, "das Logische"), "to exhibit the realm of thought philosophically, that is, in its own immanent activity or what is the same, in its necessary development". This is the idea of *the dynamic nature of forms, or, what amounts to the same, of the "dynamics" introduced into the forms (fixed in Aristotle and in the handbooks of Hegel's times) by philosophy*, i.e. by a reflexive and critical consideration about them.

2. Hegel also remarks, again in the Preface to the second edition of the *Science of Logic*, that *the forms of thought are the expression of the peculiar essence and substance of individual things:*

> If from given determinations others are inferred, it is [rightly] held that what is inferred is not something external and alien to the object, but rather that it belongs to the object itself, that to the thought there is a correspondent being. (Hegel Werke 5, 45/Hegel 1969, 50f.)

3. In the same Preface he writes:

> [I]f the nature, the peculiar essence, that which is genuinely permanent and substantial in the complexity and contingency of appearance and fleeting manifestation, is *the concept of the thing*, the *immanent universal* [my emphasis], and [if] each human individual though infinitely singular has the most fundamental of all his singularities in being a man, exactly like each individual animal has it in being an animal: if this is true, then it would be impossible to say what such an individual could still be if this foundation were removed, no matter how richly endowed the individual might be with other predicates, if, that is, this

> foundation can equally be called a predicate like the others. (Hegel Werke 5, 26/Hegel 1969, 36 f.)

It should be noted that Hegel addresses in this passage the question of the formality of logic focusing on the function of predicates in definitions. What he is stating here concerns the *conceptual (predicative) nature of logical forms* that we have discussed in Part I. Predicates stand for properties, such as "being a man", "being an animal", and are for Hegel the very conditions of expressing things and their singularity. Similarly, in the Introduction to the *Science of Logic* Hegel claims, as we saw already in Chapter 1., that logic assumes that "the determinations contained in definitions [...] are determinations of the object, constituting its innermost essence and its very own nature".

4. *Logical forms*, insofar as they express both the essence of things and the concept that makes our knowledge of things possible, *have an alethic, i.e. truth-implying, nature*. That is, they are conditions of our grasping things, and thinking truthfully. In the Preface to the second edition of the *Science of Logic* Hegel points to the generalised "scorn" of logic typical of his times.

> [Everyday thought has] so much lost its respect for the school which claims possession of such laws of truth [the law of identity and the law of contradiction] that it ridicules it and its laws and regards anyone as insufferable who can utter truths in accordance to such laws: the plant is – a plant, science is – science. (Hegel Werke 5, 28/Hegel 1969 38)

Hegel also recalls a further reason of complaint. The inference rules "quite as well serve impartially error and sophistry and [...] however truth may be defined, they cannot serve higher, for example, religious truth [...] they concern only correctness [...] and not truth" (Hegel Werke 5, 29/Hegel 1969, 38). Hegel's conception of *Wahrheit* (truth properly speaking) and *Richtigkeit* (correctness) will be examined later. Now we can see that philosophical (and religious) truth goes beyond the idea of correctness established by *Verstandeslogik*, i.e. a logic that does not critically reflects upon its forms. Such truth has different conditions and requires different forms.

5. The connection of forms and truth so introduces a *criticism of "formal logic"* as it is usually intended. In the Introduction to the *Science of Logic* Hegel criticises the view of logic typical of his times. He writes:

> When logic is taken as the science of thinking in general, it is understood that this thinking constitutes the *mere form* of a cognition, that logic abstracts from all *content* and that the so called second *constituent* of a cognition, namely its *matter*, must come from elsewhere; and that since this matter is absolutely independent of logic, this latter can provide only the formal conditions of true knowledge [original German wahrhafter Erkenntnis/original

translation: genuine cognition] and cannot in its own self contain any real truth, nor even be the *pathway* to real truth because just that which is essential in truth, its content, lies outside logic. But [...] it is quite inept to say that logic abstracts from all *content*, that it teaches only the rules of thinking without any reference to what is thought or without being able to consider its nature. For as thinking and the rules of thinking are supposed to be the subject matter of logic, these directly constitute its peculiar content; in them, logic has that second constituent, a matter, about the nature of which it is concerned. (Hegel Werke 5, 36/Hegel 1969, 43 f.)

Hegel discusses here what I have called the formalistic argument (FA) that I have already analysed.[29] In particular, he discusses the inference from the claim that logic studies the most general forms of thought, obtained abstracting away from particular contents, to the theses that logic has no content and therefore "cannot contain truth".

5.2 Against formal logic?

Some authors interpret these and similar passages as statements against formal logic,[30] and infer from them that Hegel's logic is not a formal logic.[31] But this reading risks being misleading. What I suggest instead is that Hegel does not discuss the formal nature of logic, but rather the philosophy of logic in his times, i.e. the interpretation that postulates the "abstract" (separate) nature of logical

---

29 Litt 1961, 261 criticises the formalistic approach in terms that are reminiscent of the Hegelian ones: "whether it is possible to separate form and matter in this way is not only a question that needs to be answered in a clear and objective manner; it is also and above all a *logical* question. If logic considers itself in these [formalistic] terms, then it makes the cardinal mistake to dispose once and for all of a question it should explicitly deal with".

30 See for example Krohn 1972, 107 who interprets what Hegel says on *the theories on logic* of his times as statements on formal and Aristotelian logic. For Krohn 1972, 57 Hegel's attitude toward formal logic, i.e. the logic as theory to be found in the handbooks of Hegel's times, is ambivalent. In my view, Hegel's attitude is expression of his new concept of "form" and logic's formality. Forms are for Hegel self-revising structures, and formal logic is a useful discipline insofar as it gives an account of the self-revising nature of logical forms.

31 Mittelstraß (ed.) 1980 ff., vol. 2, 57 ff. According to Ritter/Gründer/Gabriel (eds.) 1971 ff., vol. 5, 358: "[Hegel's *Wissenschaft der Logik*] radically rejects formal logic and substitutes it with a dialectic that is the product of speculative metaphysics". Butler 2012, 81 writes "dialectical logic is not a formal system of axioms with rules of inference abstracted from any particular content". However, he also states that Hegel's dialectic is "applied formal logic about a particular content".

forms.³² In other words, the theses 1.–4. considered above, i.e.: the dynamic, ontological, conceptual, and truth-implying conception of forms introduce the last aspect, the criticism of the inference from "logic is interested in the form of sentences and arguments" to "forms have no content, and have nothing to do with truth, i.e. with the relation of thought to content".

Logic, for Hegel, *is* formal in the sense of interested in the underlying structure of sentences and arguments that is obtained by abstracting away from their particular contents; but this does not mean that logic does not express real truth. Instead, the "living" nature of forms (of any kind of form) is what makes them our guide to truth. Thus, what is wrong is not formal logic in itself, but rather the way in which philosophers may think about it: "If logic is supposed to lack content, then the fault does not lie with [logic's] subject matter but solely *with the way in which this subject matter is conceived* [my emphasis]". What is wrong is hence the formalistic conception, as expressed by FA and by other common postulates of the intellectualistic and formalistic conceptions of forms.

First, in the *Logic* of the *Encyclopaedia* Hegel stresses, against the formalistic conception, that the specific subject matter of logic is a particular kind of thought, different from what we ordinarily mean by the word "thought". From a common point of view we call "thoughts" those thoughts whose content is empirical; in logic, we examine those thoughts whose content is itself thought. These are, for Hegel who on this follows Kant's terminology, pure thoughts.

> In [...] the ordinary sense we [...] intend by [thought] something that is thought of, but which has an empirical content. In logic,³³ thoughts are grasped in such a way that they have no content other than one that belongs to thinking itself, and that is brought forth by thinking. So these thoughts are pure thoughts. (Hegel Werke 8, 84/Hegel 1991, 58)

Logic's content is not thought as such, but "thought about thought", and this reflection is what constitutes forms. Forms are created by thought's natural possibility to reflect upon itself.³⁴

---

32 In this respect, Nuzzo 1997, 52 remarks that Hegel argues against the notion of formality as abstractedness, and proposes a more complex notion of formality, linked to a dialectical view of "form". For Litt 1961, 269 formal traditional logic is concerned with thought's determination of objects, dialectical logic, in contrast, could be called a logic of thought's self-awareness (*die Logik des selbstbesinnlichen Denkens*).
33 The English translation is "*In the Logic* thoughts are grasped...", which is misleading, in my view. Hegel writes in this passage about logic in general, while the English translation seems to suggest that he is referring to a particular (his own) logic.
34 Dummett 1991 and Id. 1993 considers Frege's logic a "philosophy of thought" and as such in fundamental continuity with the idealistic-Platonic view of philosophy as "thought about

Second, Hegel also says that FA is connected to a misleading metaphysics, i.e. to reductive views about "content", "matter" and "reality", according to which "content", "matter" and "reality" are supposedly external, or extra-logical. In the Introduction to the *Science of Logic* he writes:

> What is commonly understood by logic [...] has, it must be admitted, no content of a kind which the ordinary consciousness would regard as reality [...] but it is not for this reason a formal science lacking significant truth. In any event, the field of truth is not to be sought in that matter which is missed in logic and to whose lack it is custom to attribute the deficiency of logic [...] *The absence of content of logical forms is due solely to the way in which they are considered and dealt with* [my emphasis]. (Hegel Werke 5, 41/Hegel 1969, 48)[35]

"The way in which logic's subject matter is conceived" is the metaphysics and epistemology of logic underlying the formalistic account, according to which forms are rigid determinations, and contents something extra-logical.

Third, Hegel observes that:

> When they are taken as fixed determinations and consequently in their separation from each other and not as held together in an organic unity, then [forms] are dead forms. (Hegel Werke 5, 41/Hegel 1969, 48)

In this sense, critically thinking about forms implies considering their relations to each other, and the reflection upon logic is what forces us to ask about their ability to express contents. For example, asking (as Hegel claims one should do, in logic): "is the form of disjunctive syllogism true (i.e. able to convey valid and sound arguments)?" forces us to consider the relation of this form to its particular contents, namely the specific arguments that follow the disjunctive syllogism pattern ("the light is on or off, it's not on, therefore it's off"; "either you like Franz Ferdinand or Arctic Monkeys, you don't like Franz Ferdinand therefore you like Arctic Monkeys" etc.). In so doing, we are able to see when the disjunctive syllogism works, and when it does not.

---

thought". On the continuity between Hegel's and Frege's view on thought see d'Agostini 2003, 59–94.

35 I have slightly changed the 1969 translation, keeping it closer to the German text: I have translated *als Realität* as "as reality" (while the English text is "as a reality") and "Sondern das Gehaltlose der logischen Formen liegt vielmehr alleine" as "The absence of content of logical forms is due solely" (while the English translation is "The truth is rather that the insubstantial nature of logical forms is due solely"). In the English translation, Hegel's specific reference to the relation between logical forms and content is lost, and substituted by the different (and vaguer) idea of forms' substantial nature.

Fourth, in the Preface to the so called "Subjective Logic", the third part of Hegel's *Science of Logic*, which deals with topics traditionally considered in the handbooks on logic of Hegel's times, Hegel underlines that the traditional, old logic deals with rigid, ossified materials that need to be rendered fluid again. He explains that while the first two parts of his *Science of Logic* deal with completely new materials, which have never been considered before, the subjective logic presents the difficulty of dealing with a "ready-made and solidified, one may say, ossified material [...] and the problem is to render this material fluid" (Hegel Werke 6, 243/Hegel 1969, 575). The task of Hegel's logic with respect to formal traditional logic is thus *not to "build a new city in a wasteland" but "to remodel an ancient city, solidly built, and maintained in continuous possession and occupation"* (Hegel Werke 6, 243/Hegel 1969, 575). The metaphor shows that Hegel's logic is not a *new* logic, alternative to the customary one, but rather maintains the old one, and, critically reflecting upon it, re-models it. Again, the reference is here to the fact that Hegel's new logic contains a *philosophy*, a reflection on traditional logic, a reflection that implies showing the relations of the forms with each others, and to truth, and corresponds to rendering the field of logic fluid and flexible.

In sum, we can confirm the idea presented at the end of the first part: that Hegel's logic is simply what can be called a *philosophical logic*. What we can clearly see is that the contact of "logic" and "philosophy" does not imply, in Hegel's view, any dismissal of the formal nature of logic, and of formalism. It instead implies a different notion of "form", a notion that is conceived in open contrast to the formalistic and intellectualistic idea of his times, and which differs, in some interesting respects, from the most usual conception of logical form in contemporary times. What we should ask now is thus whether Hegel's ideas might entail improvements for our vision of forms.

6 Is Hegel's logic formal?

Even if discussions are still open as to what exactly the meaning of "formal" in the expression "formal logic" amounts to,[36] a preliminary assessment might be proposed.

6.1 Formalising Hegel's logic?

"Formal", as we have seen, should be distinguished from "formalised" or "symbolic".[37] In contemporary philosophical logic the formal nature of logic goes hand in hand with its formalised character, while Hegel, as we have seen, explicitly criticises the attempts, made by Leibniz and Euler among others, of using symbols to express conceptual contents. Hence, if one wishes to assess the practicability of a Hegelian approach in philosophical logic, a first question concerning logical forms and their expression needs to be asked, namely:

> "Is the dialectical conception of forms compatible with formalisation?"

Isn't a formalisation of Hegel's dialectics a needless enterprise, given that Hegel, as we have seen, criticised the attempts developed by Euler, Lambert, Leibniz among others of representing concepts using algebraic figures and signs? Isn't a formalisation a mere repetition of what Hegel already explained, and, in the worst case, a mere translation of Hegel's mistakes, in a different notation?

As we have seen, and as Hegel himself suggests, the attempts at individuating the general behaviour of our thinking and reasoning, i.e.: logic traditionally intended, is not useless at all, but rather a necessary step towards *Vernunftlogik*, to Hegel's own philosophical logic. What Hegel suggests, in his critique of symbolisms, is that mathematical symbols fix conceptual contents, while concepts are essentially fluid. However, this does not prevent him from hinting at natural, and even mathematical language as indeed apt for expressing conceptual con-

---

[36] See Dutilh Novaes 2011, 303–332 and Mac Farlane 2000 among others. Neither of them takes Hegel's reflections on the formality of logic into account.
[37] See Read 1995, 61 on the fact that the use of "formal" as synonym for "symbolic" is inappropriate. For Sainsbury 2001 chapter 6. (analogously to Russell, see here 19.2.) formalisation goes hand in hand with "extracting the forms" or uncovering the hidden structure of natural reasoning, analysing it. See Sainsbury 2001, 348 ff. for a discussion of those views (e.g. Davidson's) according to which the project of formalisation is to be distinguished from the one of conceptual analysis.

tents. Thus it is legitimate to argue that a formalisation of dialectics is Hegelian in spirit, provided that it fulfills the following two requirements: a) it is intended as an account of the most general patterns and behaviour of dialectical conceptual relations b) it is carried on in the critical and sceptical spirit that is typical for philosophy as a dialectical and speculative science.[38]

In this sense, the question: "is Hegel's logic formal or not?" can be answered affirmatively. Individuating dialectical regularities in the semantic behaviour of conceptual determinations does correspond to a "formal" analysis, and dialectic is for Hegel the form of conceptual analysis. Hegel himself also considers dialectics as a method that has general and regular traits, for example in the *Logic* of the *Encyclopaedia* (§§ 79–82) when he describes the "general structure of *das Logische* as a rule of every proceeding of the idea" (Hegel Werke 8, 168ff./ Hegel 1991, 125ff.).[39] However, as we have seen, the use of "form" has in Hegel specific connotations, possibly derived from the concept of *morphé*, which Goethe stressed in his naturalistic inquiries as a generative principle.[40] But the question arises: Is modern logic also formal in this sense? In a sense, it is. Frege was perfectly aware of this generative notion of form, when he wrote that "theorems stay within axioms not like a beam in a house, but like a plant in the seed".[41]

---

[38] Apostel 1979, 90 observes that contemporary, non-classical logics are "Hegelian in spirit", since they more or less explicitly share the assumption that language develops in force of the interaction between the common substrate (natural language) and an indefinite multiplicity of signs specially developed. In this respect they embody the very spirit of dialectics. In so doing, non-classical logics "apply dialectics to the problem of symbolism" Ibid. By this, Apostel means that the relation between dialectical conceptual contents and their symbolic expression has to be read in dialectical terms. This is precisely what Hegel intends to emphasise, in my account.
[39] See Marconi 1979b, 16ff. and Sacchetto 1993, Chapter 13 for a discussion of the relationship between dialectic and formal logic.
[40] On Goethe's theory of living forms see Moiso 2002, Cislaghi 2008, 172f. as well as Breidbach 2006. On the organic nature of philosophical thought see Litt 1961, 18ff. Abel 2004, 325ff. highlights the dynamic nature of forms and the interplay between forms of knowledge (*Wissensformen*) and contents. On the "constitutivity" of logical form for Hegel see Gerhard 2015, 18. On Hegel's notion of form as "form that generates content" see Zambrana 2017, 299–300.
[41] See Frege 1884 (section 8). Interestingly, Hegel's criticism of Leibniz's *calculus* parallels Frege's critique of Boole's algebra. As Gabriel 2008 shows, Frege adopts many aspects of Trendelenburg's critique of Leibniz. In my view, the arguments adopted by Frege from Trendelenburg are the same Hegel adduces against the *Merkmaladdition*. One significant concept figures in all three authors: the concept of organism, and of the organic nature of conceptual relations (Gabriel 2008, 121 quotes the famous "scholar-scene" in Goethe's *Faust* as common inspiration of both Trendelenburg and Frege). What all these authors call for is the necessity of expressing the living

6.2 One hypothesis

There are now non-classical conceptions of logic that are particularly close to the Hegelian view of forms. Hegel's critique of the incapability of traditional logic to convey truth anticipates relevant logic's critique of classical logic.[42] As we have seen, Hegel writes that "the true cannot be found in these [Aristotelian logical] forms", that "the form of an inference may be absolutely correct, and yet the conclusion arrived at may be untrue", and that this is the sign that "this form as such has no truth of its own. But from this point of view" Hegel concludes "these forms have never been considered".

One could argue that now logical rules are being considered precisely from this point of view. In a more traditional perspective, that classically valid forms fail to express sound (i.e. true, relevant, fruitful, strong) arguments is simply taken to mean (as in accordance with FA considered above) that logic is merely formal, and has nothing to do with the content of thought. In contrast, in contemporary philosophical logic a more Hegelian point of view has arisen. According to it the inability of classical logical forms to convey sound arguments is sign of a *failure* of classical logical forms, the source of a critical reflection on the same forms, and of the need of revising or enlarging classical logic.[43]

Furthermore, Hegel's critique of formalisms is, as we have seen, connected to truth. Logic in a contemporary sense deals with truth (since validity is anyhow defined as truth-preservation), though not properly with the effective, realistic truth of our usual inquiries, but rather with the assumption of truth, as related to abstract domains, or to axioms, or to possible worlds. In this respect, Hegel's logic is not formal in the sense of being "uninterested in the content", but rather in the sense that it aims at individuating the valid form of philosophical, conceptual reasoning. And this is a kind of "formal" admitted by contemporary non-classical logics, such as relevant logics.

---

nature of forms and conceptual relations. In this sense the same Fregean *Begriffsschrift* arises from a need that can be traced back to Hegel: to find a language that is as compatible as possible with the organic conception of forms, i.e. to the living nature of concepts. See also Käufer 2005, 259–280.

42 For an overview on relevant logics from the point of view of the controversy on the Law of Non-Contradiction see Berto 2007a, Chapter 9.

43 In this spirit, Read 1995, 2 complains about "a widespread but regrettable attitude towards logic, one of deference and uncritical veneration. It is based on a mistaken belief that since logic deals with necessities, with how things must be, with what must follow come what may, that in consequence there can be no questioning of its basic principles". Berto 2007a, 187 writes that "the notion of relevance goes beyond the mere realm of pragmatics; it fully belongs to logic, and it can be supplied with a rigorously formal treatment".

However, Hegel's conception of logical forms also differs from the standard one in some important respects.

6.3 What are logical forms?

Logical forms are generally intended to be linguistic structures that can be repeated and on whose basis we establish the validity of arguments.[44] For example, the arguments:

> Berlin is in Germany and Berlin is in Europe, hence Berlin is in Europe

and

> Giacomo loves Silvia and Silvia is a teacher, hence Silvia is a teacher

and

> Rome is in Italy and today the sun shines, thus the sun shines today

have the same form: p and q, therefore q. The form is valid insofar as, for every substitution of the non logical terms "p" and "q", it conveys conclusions that cannot be false, given that the premises are true.

In the case of sentential forms, such as "S is P" or "for all x, x is P", we establish the truth or falsity of the sentence on the basis of the substitution of "S" and "P" or "x" and "P" (if S = cat and P = animal then we have a true sentence, if S = Donald Trump and P = Democratic candidate then we have a false sentence).

All this stated, two aspects concerning Hegel's view of logical forms mark its originality with respect to contemporary views. First, in a standard account the substitutions are given on the basis of domains.[45] In this conception, which goes back to Tarski, we fix the domain of entities that can take the place of "p" and "q" (or S and P) in our inferences and sentences and establish on that basis the truth and falsity of our sentences, and the validity or invalidity of our inferences. For Hegel the substitutions are given by the world, and by the use and meaning of words in our natural or scientific language.

---

44 See Sainsbury 2001, 44.
45 See Tarski 1999, 115–143.

Second, for Hegel the form of sentences and inferences depends on the form of concepts. The guarantee of truth and validity is given by the conceptual content, and the conceptual content is given in the definition or conceptual determination. In this light, the reason why Hegel claims that formal logic is not sufficiently formal, and that the forms are themselves content, also becomes clear. He means that the forms of inferences and sentences are to be rooted in the form of conceptual thought. Moreover, when Hegel emphasises that the conceptual forms are living and dynamic, what he means is that they are rooted in the natural logic of language.[46]

---

[46] From this point of view dialectical conceptual determinations have been interpreted in terms of conceptual stipulations, and distinguished from definitions. For Marconi 1979b, 18 ff. a mechanism of continuous re-definition of conceptual terms is at work in Hegel's dialectic. Dialectic is an examination of the conceptual terms' syntactic and semantic articulation in natural language. Such an examination and continuous re-definition is possible because natural language is indeterminate. Hence the vagueness and indeterminacy of natural language provide the reason why Hegel's dialectical processes, for Marconi, are best grasped as processes in which we propose stipulations about the meaning of the conceptual terms. For this reason dialectic is not a conceptual dictionary, in which the definitional process would come to an end. The hypothesis according to which words have a determinate meaning implies accepting the authority of a particular theory over the language and, for Marconi, Hegel refuses to accept the authority of any theory. In fact, a theory implies first that we establish the syntactic role of that term, e. g. that it can appear in the subject and not the predicate position in a sentence, and then fix the set of the possible substitutes of that term. While technical languages are bounded by theories, the terms in natural language are not, and this means for Marconi that they are susceptible of partial, or even contradictory determinations. In sum, dialectic for Marconi corresponds to the language's effort to determine itself, a process in which its indeterminacy generates contradictions. The discovery of the contradiction forces us to abandon the demand of definitive conceptual determinations, and to look for new ones. See especially Marconi 1979b, 73. In this perspective, dialectic would be a heuristic and explorative activity, an investigation into the conceptual connections implied by the use of language, connections that are neither indeterminable nor one-sidedly determined, but rather vaguely determined, in different and incompatible ways. Dialectic for Marconi 1979b, 70 rejects "the requirement of a universal codification of the discourse – it rather tries to adhere to the natural semantic determinations". See also Nuzzo 2010b, 61–82 and, for a critique of both Nuzzo and Marconi, Bordignon 2013, 179–198.

Summary

The second part is dedicated to an analysis of Hegel's concept of logical forms and formal logic, and to its possible contribution in current debates. In **Chapter 4** I consider Hegel's discussion of views of logical forms developed in the history of philosophy, in particular by Aristotle, the Stoics, Leibniz and Kant). Two aspects are important with respect to Hegel's interpretation of Aristotle: 1) Hegel's appraisal of Aristotelian logic, considered by him as a "masterwork" and the best example of intellectual logic or *Verstandeslogik*. 2) Hegel's critique of *Verstandeslogik*. On this point I propose a revision of dominant interpretations. As opposed to these, I argue that Hegel did not criticise the general project of a formal logic, but rather the *theories on formal logic* current in his time. According to such theories "the forms are only forms and have nothing to do with the content". Hegel questions this very claim. In other words, while the interpreters claim that Hegel states something like: "formal logic is useless because it does not tell us anything about the content of our thought", I stress that Hegel claims instead that "those who say that formal logic has nothing to do with the content of thought and that the forms studied by logic are empty are wrong". The forms are for Hegel, who, in this respect, follows Aristotle, expression of the essence of things. The problem with Aristotle's *Verstandeslogik* is rather the following: Aristotle does not give an account of the logic at the basis of his own logic, but simply enumerates the forms without explaining their reciprocal connections, and their origin in self-reflexive thought. Stoic logic is important for Hegel because it contributes to a more precise definition of logical forms as *structures of thought thinking being*, i.e. *forms of truth*, whereby the question about the distinction between forms of being and forms of thought, and their relation, was not explicitly addressed before. Hegel's reading of Leibniz is fundamental for my aims in this chapter insofar as it allows for an assessment of Hegel's position on formal logic intended as *formalised or symbolic* logic. I explain here that Hegel's attitude is ambiguous: on the one hand he rejects every attempt, such as the Lebnizean one, at expressing conceptual relations using symbols. On the other hand Hegel holds that *every* language, natural language included, is inadequate to express conceptual relations, and yet he uses natural language, stressing that we only have language (a symbolic instrument) in order to express conceptual dynamics. Moreover, in one early writing (the *Differenzschrift*) he does try to formalise dialectic, and in the *Science of Logic* he praises mathematical language as better able to grasp the nature of the true infinite. All this is indicative that Hegel's critique of the project of both a *lingua characteristica* and *calculus ratioci-*

nator is to be understood as a cautionary indication on how to use symbols, and not as an interdiction.

Hegel's confrontation with Kant highlights, first of all, the common basis of the transcendental-philosophical and the dialectical projects with respect to logic, namely the idea that forms are "deposited" in our common thought and even reality (Kant states "everything happens according to rules"), and that they are *norms* and *modalities* of thought. On this common basis it is possible to fix the meaning of the expression "formal logic" and of the adjective "formal" in both thinkers: formal logic emerges from reflecting on the *modalities of our thought*. Accordingly, "formal" means *reflexive*, implying a semantic ascent from thinking about something to thinking about our thinking about this something. Formal logic is a discipline involving an *abstraction* from particular contents, aiming at seeing what is common to every content. "Formal" hence also means "*general*". Formal logic brings us to discover that without which we could not think at all, the *necessary conditions* of our thought, and "formal" thus implies a connection to necessity, and has a *normative* meaning: forms are norms. All these aspects emerge in Kant's *Jäsche Logik*, and are kept within Hegel's account. The difference between the two views emerges in relation to the concept of truth and its link to logic, in the passage from the *Subjective Logic* were Hegel discusses Kant's claim on truth in the *Introduction* to the *Transcendental Logic* of the *Critique of Pure Reason*. I reconstruct Kant's argument (I call it formalistic argument, FA), and Hegel's critique. In short, FA states:

1. The concept of truth that is fundamental in logic is correspondence between knowledge and object.
2. What we are looking for (in logic) are universal criteria for the truth of all knowledge.
3. A universal criterion must be valid for all cognitions without distinction of their objects.
4. With such a criterion abstraction would be made from all content of cognition, i.e. from any relation to its object.
5. Since truth concerns precisely this content to ask for a mark of the truth of this content of cognitions is impossible and absurd.
6. Sufficient and at the same time general criteria of truth cannot possibly be given.

Hegel claims that the argument is self-contradictory. It states 1. (truth is correspondence between thought and content), and then its negation (5.: truth concerns only the content). According to Hegel, 5. is simply false, while 1. is true. Truth concerns the *correspondence* of thought (forms) and content, and not *only the content*. This means that the task of looking for universal criteria of truth is not meaningless. The point is rather, for Hegel, to work on the forms by subjecting them to a critical analysis. Moreover, for Hegel to say that looking for forms of truth is meaningless, thus relegating the question of truth to the content (which we cannot check it in each case) is *dangerous*. It hinders us from questioning the very forms established by logic.

In **Chapter 5** I re-consider Hegel's concept of logical form by taking into account Hegel's published writings, and in doing so I individuate five Hegelian theses about forms. Forms for Hegel are or should be: 1. *dynamic*, 2. *essence-revealing*; they have 3. a *conceptual basis*, and are 4. *truth-implying*. The theses 1. to 4. introduce the last aspect, Hegel's criticism concerning logical forms. Again, textual analysis shows that Hegel's critical attitude is against then dominant *views of logical forms and formal logic, and not primarily against "logical forms" and the general formal logical project*. Hegel criticises philosophers who fail to acknowledge the points 1. to 4. and intends his *Vernunftlogik* or speculative-dialectical logic as a philosophical, i.e. a self-critical and complete account of the formal-logical realm.

In the **Chapter 6** I consider the question: "is Hegel's concept of logical forms and formal logic compatible with formalisation?" and, derivatively, "is a formalisation of Hegel's dialectic Hegelian in spirit?". Both questions can be answered affirmatively. As to the first, there are conceptions of formal logic that are particularly close to the Hegelian view of forms. Second, I argue that formalising Hegel's logic is Hegelian in spirit, provided that it fits the following two requisites: a) it is intended as an account of the most general patterns and behaviour of dialectical conceptual relations and b) it is carried out in the critical and sceptical spirit that is typical for philosophy as a dialectical and speculative science. I conclude by suggesting what I see as the difference between Hegel's and commonly accepted accounts of logical form.

III Truth

The [...] explanation, [according to which truth is] the correspondence of knowledge with its object [has] great, indeed supreme, value. (Hegel Werke 6, 266/Hegel 1969, 593)

The Hegelian conception of truth is often held to be untreatable from a logical point of view, and more specifically unapproachable in standard truth-theoretic terms.

A first difficulty is that Hegel claims that sentences are not able to express conceptual (speculative, philosophical, concrete) truth.[1] This seems to compromise from the very beginning any attempt at reading his theory from a strictly logical point of view. Some authors go so far as to suggest that this claim seems to contradict Frege's view according to which the sentence is the primary logical and also ontological unity.[2] Similarly, others read the famous Hegelian statement "the true is the whole" as implying that *thus* truth cannot be the property of single sentences.[3] The further Hegelian claim that "the true is the process" and the Hegelian idea about the dynamicity of conceptual thought also seems to suggest a major incompatibility with respect to the common logical account, according to which truth, being the property of single sentences, is fundamentally static.[4] Finally, another common view is that truth, for Hegel, is not the property of sentences or propositions but rather a "property of things".[5] From this, the complaint emerges that in Hegel there is a complete lack of semantics, no awareness about the distinction between the linguistic/sentential and the ontological level.[6]

Stressing the non-sentential nature of Hegelian truth, however, does not alone give a complete account of Hegel's position, which also includes claims such as: "the sentence [*Satz*] is truth" (Hegel Werke 6, 311/Hegel 1969, 631) and "the sentence [*Satz*] is were [...] the field of truth begins" (Hegel Werke 4, 105). Besides this, in relation to the Aristotelian account he writes: "sentences (Sätze) are were affirmation (*kataphasis*) and negation (*apofasis*), were falsity

---

[1] "The form of the sentence (*Satz*) or more specifically of the judgement (*Urteil*) is not suited to express the concrete – and the true is always concrete – or the speculative. Every judgment (*Urteil*) is by its form one-sided and, to that extent, false" Hegel Werke 8, 98/Hegel 1991, 69.
[2] Tugendhat 1970, 152 writes: "Hegel shared the prejudice of the logic of his times according to which judgements are composed of concepts, and the speculative logic that he developed is a logic of the concepts and determinations and systematically violates Frege's view that the primary logical and, one could add, also ontological unity [...] is the sentence". Brandom 2014, 2f. stresses a similar point, though with a different accent, when he praises Hegel for dismissing the view according to which sentences are the basic unities of objective knowledge and truth.
[3] See the holistic coherentistic account of Joachim 1999 and more recently Brandom 2005, 131–161.
[4] Brandom 2005.
[5] See Stern 1993, 645–647 and Baldwin 1991, 35–52.
[6] See Puntel 2005, 208–242.

(*pseudos*) and truth (*aletheia*) happen" (Hegel Werke 19, 235/Hegel 1892 ff., vol. 2, 217).

A second difficulty concerns the very meaning of the predicate (or conceptual function) we call "truth". As mentioned, Hegel claims that the meaning of truth as correspondence is "of supreme value".[7] At the same time, however, he writes about "correspondence of the object with itself" or "of the object with its concept" or "of the concept with itself". Moreover, his view about the link between reality and rationality, which we may take to be the two terms of the traditional correspondentistic conception, is paradigmatically expressed by the double sentence "what is rational is real and what is real is rational". This has been taken to involve a kind of collapse between rationality and reality, and the view that reality is constituted by, and not independent from, rationality.[8]

All this has led most interpreters to think that by "correspondence" Hegel actually meant something else. The result is a somewhat puzzling account of Hegelian truth as, by turns, correspondence between thought and thought, and thus hardly distinguishable from coherence,[9] correspondence between being

---

7 Stekeler-Weithofer 1992, 34f. shows that for Hegel correspondence is the fundamental and inevitable meaning of truth. The point is rather, for Stekeler-Weithofer, to ask: "What corresponds to what, in Hegel?". Stekeler-Weithofer 1992, 35 also claims that, in order to give an adequate account of Hegel's truth theory, both elements, the constructivist and the factual-real one, are to be considered.
8 See also in general the interpretations of Hegel's idealism in anti-realistic terms, for example Sprigge 2002, 225 f.
9 See the classical development of Hegelian themes in the holistic coherentistic account of Joachim 1999, 50 ff., as well as the holistic inferentialistic account, in which a coherentistic account of truth seems to be a basic assumption. Lau 2004, 35 ff. argues that Hegel's philosophy does imply coherentism about truth (see 61). However, for him coherentism is only one aspect of the Hegelian perspective, which also includes correspondentism. "The coherence with the system and the correspondence to reality as correspondence to its concept are one and the same" (Lau 2004, 62). The authors who stress Hegel's coherentism classically deny that he rejected the Law of Non Contradiction. Stekeler-Weithofer 1992, 348 and 420 highlights the coherentistic and pragmatic component in Hegel's concept of truth. Hegel's *Habilitationsthese* "*contradictio est regula veri...*" means, for Stekeler-Weithofer, that the emergence of inconsistencies is the sign that something in the communication went wrong, and needs to be corrected. Hösle (1998, 173) writes: "the finite that is not ideal moment of the infinite is contradictory; the infinite that is opposed to the finite is contradictory. The true infinite, in contrast, which is the unity of finitude and infinity, is free from contradictions" (the passage is quoted by Lau 2004 at 61). Also for Berto Hegel's dialectics does not imply any challenge to the Law of Non-Contradiction. Berto (2005, 67) stresses, similarly to Lau, that coherentism and correspondentism do not exclude each other: for Hegel truth is correspondence between knowledge and object, whereby the object is constructed by knowledge. This means for Berto that Hegelian truth is *adaequatio* between

and being, and thus actually not "truth" anymore (Puntel 2005, 208–242) neither correspondence nor coherence but *identity* between concept and object or thought and reality,[10] a combination of correspondentism and coherentism (Berto 2005, 67 and Lau 2004, 52).

These interpretations, however, are not able to give an account of other Hegelian statements, such as, among others: "the [...] explanation, [according to which truth is] the correspondence of knowledge with its object [has] great, indeed supreme, value" or "truth is the correspondence of my knowledge with the object" (Hegel Werke 4, 291) and "reality [is] what comes first and [what] to which the concept must correspond in order to be true" (Hegel Werke 4, 203).

In this panorama of conflicting accounts, a new perspective might be profitable. More specifically, the point of such a perspective would be to clarify how and why Hegel criticised the sentential form and yet declared sentences to be the only way we have to express truth, and how and why Hegel defended the classical correspondentistic conception, while rejecting the static view of the relation between thought and being.

My analysis shares the spirit of contemporary works interested in showing the relevance of Hegel's philosophy for analytic philosophy and logic (see Berto 2005, Brandom 2002, 2005 and 2014, Koch 2014, Nuzzo 2010a, Pinkard 2003, 119–134, Pippin 2016, Redding 2007 and 2014, 281–301, Stekeler Weithofer 1992, 2005 and 2016, 3–16). Even if the main focus of these researches is not primarily Hegel's notion of truth, they are important for clarifying the actual relevance of Hegel's concept of *Begriff* (the conceptual realm). In particular, Brandom's analysis (Brandom 2014, 1–15) highlights two aspects of Hegel's notion of the conceptual realm, its normative and its inferential dimension.[11] I share

---

intellectus et intellectus rather than between *intellectus et rei*. A problem with this account is that Hegel never states that the object of our knowledge is *constructed* by our cognitive functions. Differently from Berto and Lau, I stress that, insofar as the question about the definition of truth is concerned, Hegelian truth is to be read in classical correspondentistic terms.

10 See Lau 2004 and, in a different perspective, Baldwin 1991, 35–52. For a complete assessment of the question: "did Hegel hold an identity theory of truth?" see Miolli 2016. For Miolli 2016, 21f. Hegel's view is different from the so called "identity theory of truth" insofar as it implies a more complex vision of truth as "thought about the thing" ("pensiero della cosa") whereby this expression has a double meaning: first it means that, for Hegel, thought is objective and second that "the thing" is reality mediated through thought.

11 On the first point, Brandom claims that conceptual determinations and descriptions are never totally free from normative and prescriptive commitments, and that normative commitments have to be considered in one with the descriptive ones. As Brandom 2014, 10 puts it: "here is how I think the social division of conceptual labour understood according to the recognitive model of reciprocal authority and responsibility works in the paradigmatic linguistic case,

the view about the normative and inferential nature of the conceptual realm, but I stress that this view does not prevent Hegel to intend truth classically, in realistic and correspondentistic terms.¹²

In what follows I keep myself to what Hegel says about truth. My analysis is oriented by two standard questions at the basis of every truth theory: what are the truth-bearers for Hegel?, and what does the word "true" mean, for him? Focusing on some textual passages from the "Subjective Logic" (its version in the *Science of Logic*, in the *Lectures on Logic and Metaphysics*, in Hegel's *Nürnberger Schriften* and in the *Encyclopaedia*), as well as on the *Lectures on the History of Philosophy*, I first examine Hegel's view on truth-bearers (in Chapter 7.), and second his statements on the meaning of "true" (in Chapter 8.). In Chapter 9. I assess the role of Hegel's view within contemporary debates on truth.

---

so as to resolve the tension between authority over force and authority over content. It *is* up to me which counter in the game I play, which move I make, which word I use. But it is *not* then in the same sense up to me what the significance of that counter is – what other moves it precludes or makes necessary [...]". The essays collected in Halbig/Quante/Siep (eds.) 2004 are proof of the central role played by normativity in the Hegel reception of the last 15 years. On the normative character of Hegel's philosophy in general and of his concept of spirit in particular see Pinkard 2002, part III. On the second aspect Brandom 2014, 1 stresses that Hegel's view of the conceptual realm was anticipated by Kant's insight according to which "particular and general representations, intuitions and concepts are to be understood only in terms of the functional role they play in judgements". In this respect, Hegel further develops Kant's approach not only understanding "concepts and objects in terms of judgements, but judgements in terms of their role in inference" (Brandom 2014, 2). On truth in Brandom and Hegel see Ficara 2020, 29–40.

12 The theme of the relation between the conceptual realm and truth is the subject of Pinkard 2003, 119–134. Pinkard stresses the normative component of Hegel's conceptual truth (*Wahrheit*). He claims that to grasp the meaning of Hegel's concept of truth we must understand what it means that concepts *realise* themselves. For Pinkard "concepts are realised insofar as we act following them", the reality of the concept is the way in which the concept works as a normative instance. In this respect Pinkard highlights the practical consequences of Hegel's view on conceptual truth. He shows that concepts for Hegel are teleological structures, which force us to follow them, think, act and live according to them (Pinkard 2003, 121). Also Brandom 2005, 131–161 addresses the question about the meaning of truth at the basis of Hegel's account of the conceptual. According to Brandom, Hegel's truth is (rightly, in my view) the whole process through which we inferentially determine the content of concepts, and transform judgements. However, this implies for Brandom that truth, "Hegelianly" intended, is not primarily a property of single judgements, and is to be intended in pragmatic and non-realistic terms. Truth is, as Brandom stresses, "something we make" rather than "something we have". I emphasize, in contrast, that truth is, in Hegel's view, both the process through which we inferentially determine the content of concepts, *and* a predicate (or predicative function) we use to express the property of propositions, claims, statements, assertions, or any other truth-bearer. Moreover, I stress that truth, for Hegel, is both something we make *and* something we have.

7 Truth-bearers

In the German pre-Hegelian logical terminology two terms: *Satz* ("sentence") and *Urteil* ("judgement") are used by several authors to express the Aristotelian *logos apophantikós*, the logico-linguistic unity that can be true or false.[13] "*Satz*" and "*Urteil*" both refer, in the Hegelian terminology, to the sentential expression of a thought (which, in the Aristotelian logic of Hegel's times, consists in a subject and a predicate joined by the copula), equivalent to the Aristotelian *logos apophantikos*.[14] In the *Phenomenology of Spirit* Hegel admits, beside normal "Sätze/sentences" also what he calls "speculative sentence". The latter is the genuine *logos apophantikos*, i.e. the truth-conveying *logos*, while the former are not apt to express conceptual truth.[15] As we will see, in his later writings Hegel distinguishes between *Urteil* (the conceptual, speculative, genuinely truth-bearing sentence) and *Satz* (the non-conceptual sentence). Here he claims that while every *Urteil* is a *Satz*, not every *Satz* is a *Urteil*. In this respect the dis-

---

[13] In his German Logic Wolff translates the Latin "*iudicium*" as "*Urteil*" in the sense of "*logos apofantikós*" while *Satz* is the translation of the Latin *propositio* and has the meaning of "Setzung", i.e. position. Kant uses *Satz* in the *Jäsche Logik* for expressing the assertive, and *Urteil* for the problematic sentence. This use is reversed by Frege in *Funktion und Begriff* (now in Frege 2008, 1–22) who calls *Satz* the mere position of a case in an assumption, and *Urteil* the assertoric statement. In 1893, in contrast, Frege calls *Satz* the "begriffschriftliche" representation of an assertoric judgement. See on this Ritter/Gabriel/Gründer (eds.), vol. 8, 1193. According to Ritter/Gabriel/Gründer (eds.), vol. 11, 436 the different uses and linguistic traditions go back to different views on the (linguistic, psychological, ontological) nature of the bearer of truth, as well as to different ways of intending the meaning of the copula in the sentential form "S is P".

[14] In contemporary English the expression that is closest to Aristotle's *logos apophantikos* is, possibly, "declarative sentence". The term "proposition" has a more specific meaning, and stands for the content of a declarative sentence. Thus in the following pages I use "sentence" for "*Satz*" and "judgement" for "*Urteil*".

[15] Schäfer 2001, 158 ff. and 194 f. explains that in Hegel's conception of the speculative sentence in the *Phenomenology of Spirit* are already contained the essential aspects that constitute Hegel's mature conception of dialectic. Dialectic is here the "method corresponding to the self-movement of speculative determinations". As Schäfer shows, see also Düsing 1976, 198 ff., in the *Science of Logic* the representation of dialectic through the speculative sentence is substituted by its inferential, syllogistic articulation. However, the important insight at the core of Hegel's mature dialectical logic is already given in Hegel's *Phenomenology of Spirit*, and is what Schäfer calls the "unity of method and content". I spell it out as the descriptive and normative nature of dialectical logic. In dialectical logic we analyse conceptual determinations, the leading words and views that orient our thought and action and make it possible. We observe their nature, they show themselves to be dialectical, i.e. they turn into their opposites, and their dialectical nature becomes the norm according to which thought must proceed in its search for truth.

tinction recalls his early theory of the "speculative sentence [Satz]".[16] In what follows I do not present the development of Hegel's theory of judgements from the early to the late works. Rather, I refer mainly to Hegel's general definition of *Urteile* in the "Subjective Logic" in the *Science of Logic* by focusing on the link between the sentential form (as *Satz* and *Urteil*) and truth.

7.1 Hegel and the sentential nature of truth

Judgements are, in the logical tradition Hegel refers to, dealt with in the so called *Urteilslogik*, the "logic of judgements", which comes immediately after the *Begriffslogik*, the "logic of concepts". Some authors claim that the use of "*Urteil*" has a psychological connotation. They consider Kant's definition in the *Jäsche Logik* ("a judgement is the representation of the unity of the consciousness of different representations") as illustrative of the psychological perspective.[17] The corresponding logico-linguistic expression would be "declarative sentence" (*Aussagesatz*) or simply "sentence", and the ontological one *Gedanke*, or "proposition" as content of a declarative sentence.[18] However, even if historically Hegel's logic does belong to the so called "psychological" period, there is no trace of a psychological sense of "judgement" (*Urteil*) in Hegel. In the *Encyclopaedia Logic*, we read:

> The judgement is usually taken in the *subjective* meaning, as an *operation* or form, which only emerges in *self-conscious* thought. This distinction is not present in the logical field, the judgement should be taken in completely general terms: *All things are sentences* – i.e. they are *individuals*, which are in themselves a *universality* [...] or they are a *universal*, which is *individualised*. The universality and the individuality is distinguished in them, but is at the same time identical. (Hegel Werke 8, 318 f./Hegel 1991, 245 f.)

Hegel explains that the view according to which judgements are operations or processes of thought is not the one generally presupposed in logic. According

---

[16] See on Hegel's view on sentences/judgements and its development in Hegel's writings Bodammer 1969, Düsing 1976, Lau 2004, Campogiani 2006.
[17] However, an interpretation of Kant's view on judgements and of Kant's logic in general in psychologistic terms would be misleading. In the *Jäsche Logik* Kant explicitly stresses the difference between the psychological consideration, which deals with *how we think*, and the logical one, which is focused on *how we should think*. In the *Critique of Pure Reason* Kant also hints at the limits of the traditional definition of the judgement as "representation of the relation between two concepts". Kant B 140, see further on this Tugendhat/Wolf 1993, 18.
[18] Tugendhat/Wolf 1993, 17.

to him the strictly logical meaning of "judgement" is rooted in the correspondence between the structure of reality and the structure of sentences. The quote shows the Aristotelian background of Hegel's view on judgements. Hegel says that judgements express a relation between universals and individuals, and that this relation is typical of reality itself. As it emerges from the quote, according to Hegel reality is structured in terms of what he calls "things", and Aristotle calls "substances" (*ousiai*). This relation of instantiation is expressed in a sentence through the relation between the term that stands for the individual, i.e. a name or definite description, the subject, and the one that stands for the property or universal, i.e. the predicate. In this sense, interpreting Hegel's standpoint on judgements in psychologistic and pre- or non-Aristotelian terms would be misleading.[19]

For Hegel the judgement is thus, clearly, *logos apophantikos* in the Aristotelian, and also Platonic, meaning. In the *Sophist* Plato observes that the *logos*, intended as the union of a name and a predicate, is the minimal unity through which we communicate and show something. In *De Interpretatione* Aristotle refers to the same notion of *logos* presented in the *Sophist* as *logos apophantikos* and defines it as the linguistic unity which can be true or false.[20] Hegel substantially shares this view when he states that the "the judgement is truth" (Hegel, Werke 6, 311/Hegel 1969, 631) and "the sentence is where [...] the field of truth begins" (Hegel Werke 4, 105).

7.2 Hegel's critique of the sentential form

Hegel's view on sentences also entails a critique of the sentential form and its capability to express truth. On Aristotle's *De Interpretatione* in his *Lectures on the History of Philosophy* Hegel writes:

> [sentences/Sätze] are not were *nous* thinks itself and is in pure thought; [they are] not universal, but rather particular [predication]. (Hegel Werke 19, 235/Hegel 1892ff., vol. 2, 217)

I have mentioned in Part II. that pure thought is for Hegel reflexive thought, "the thought about thought". Now we see that when *nous* (thought) thinks itself and we find ourselves "in the element of pure thought" or "universality" the limits of the sentential form emerge. Hegel is referring here to the Aristotelian concept of

---

[19] See Lau 2004, 56, who confirms my point.
[20] See Plato *Sophistes*, 262 c-d (translation from Hamilton/Cairns 1961), Aristotle *De Interpretatione*, Chapter 4 (translation from Barnes 1984) and Tugendhat/Wolf 1993, 21ff.

nous as "thought thinking about itself", thought considered in itself, having only thought as an object, without any empirical substrate. The reference to Hegel's own perspective in the *Science of Logic* is evident. For Hegel logic deals precisely with the field of pure thought, and is the scientific development of thought thinking about its own structures or forms. As we have seen, the forms of thought isolated and studied by logic are for Hegel, as well as in the Platonic and Aristotelian tradition, the expression of the essence of things, and of truth.

When it comes to express the self-referential and pure nature of thought the limits of the sentential form emerge. In § 31 of the *Encyclopaedia* Hegel writes:

> the form of the sentence (*Satz*) or more specifically of the judgement (*Urteil*) is not suited to express the concrete – and the true is always concrete – or the speculative. Every judgment (*Urteil*) is by its form one-sided and, to that extent, false. (Hegel Werke 8, 98/Hegel 1991, 69)

And in the *Lectures on the History of Philosophy*, he says that "sentences are not when the nous thinks itself", and that "the speculative cannot be expressed as sentence" (Hegel Werke 19, 397/Hegel 1892ff., vol. 2, 369 – see also 364). Single sentences are not able to express "the concrete", or "the speculative", which is one and the same as "the true" or "philosophical truth".

So we are facing the already mentioned dissonance. How can we then reconcile the basic idea that sentences are the genuine truth-bearers, with the idea that logical truth does not have sentential/propositional nature?

7.3 "The true is the whole"/"The true is the process"

The roots of Hegel's critique of the sentential form can be found in the *Phenomenology of Spirit*, in Hegel's claim that:

> The true is the whole. The whole, however, is merely the essence that completes itself through the process of its own development. (Hegel Werke 3, 24/Hegel 1977, 11)

According to Nuzzo, in the preface to the *Phenomenology of Spirit* Hegel lays out a conception of truth that is the very frame of the *Science of Logic* and, more generally, of his "mature" view on logic.[21] In accordance with this preliminary insight, I suggest that the *whole/process* claim ("the true is the whole"/"the true is the process") can be read as an answer to the question: "what are the genuine truth-bearers, for Hegel?", but without modifying, substantially, the basic idea

---

21 See on this Nuzzo 2011, 91–105.

that for Hegel the truth-predicate has a propositional nature: it is a predicate applied to sentences/propositions.

– Of "the whole" Hegel writes that:

> The true form in which truth exists can only be its scientific system. (Hegel Werke 3, 14/Hegel 1977, 4)

And in a successive passage he specifies:

> Identifying the true form of truth with this scientific character [Wissenschaftlichkeit i. e. scientificity] is the same as stating that truth finds the element of its existence in the concept alone. (Hegel Werke 3, 14f./Hegel 1977, 4)

Hegel refers here to the scientific system, which he identifies with "the concept", as the only form or element in which "truth can exist".

What Hegel means is revealed in the following passages of the preface, where he explains the difference between the ancient and modern philosophers' attitude towards the concept or universal – Hegel also uses the plural "concepts" and "universals", and calls them "pure essences". The ancient philosophers discovered universals by "extracting" them from our concrete experience or, we could say: from our natural logic and metaphysics. The moderns, by contrast, had them already there. Their contribution rather consisted of distinguishing them from one another, and fixing them.

On Hegel's view scientific truth is thus always relative to the concept or universal. As we will see better later, this does not mean that the concept, rather than the sentence/judgement, is the truth-bearer for Hegel. Rather, it corresponds to the idea, which is also typical of the Aristotelian concept of *episteme*, that scientific knowledge involves totality and universality, i.e. grasping *everything* there is to know about something.

– Now, regarding truth as "process" the problem is for Hegel to "bring fluidity to fixed thoughts" to "spiritualize the universal through the overcoming of fixed, determinate thoughts" (Hegel Werke 3, 37/Hegel 1977, 19f.). Hegel uses here a new word, "*begeisten*", different from but reminiscent of *begeistern*, which means to enthuse, inspire. *Begeisten* literally means "to introduce spirit", or "spiritualize".

Thus "*spiritualizing the universals*" *means introducing movement into rigid intellectual categories.* As Hegel writes:

> Through this movement pure thoughts become *concepts* and become what they are in truth: self-movements [...] spiritual essences [...] this movement of pure essences corresponds to the nature of the scientific method [*Wissenschaftlichkeit*, scientificity]. If we consider [the

movement] as the connection of their content, [the movement] is the necessity and the display of the content into the organic whole. (Hegel Werke 3, 37f./Hegel 1977, 20)²²

The second part of the whole/process statement is also to be understood in this light, namely:

> The whole is merely the essence that completes itself through the process of its own development. (Hegel Werke 3, 24/Hegel 1977, 11)

The whole is the "essence that completes itself" means that what something is, i.e. the concept or essence of something, what something really is, undergoes a development, and that that development is "the whole" which can be said to be true, to be the bearer of truth. From this point of view the Hegelian picture implies that scientific (total, universal, conceptual) knowledge is not fixed once and for all, but undergoes developments and changes in history.

– *The reflexive nature of truth*. The "movement" of the concept is due to the reflection that constitutes the specific prerequisite of *Vernunftlogik*. In this respect, Hegel stresses that "truth is not a minted coin":

> Truth and falsehood as commonly understood belong to those sharply defined ideas which claim a completely fixed nature of their own, one standing in solid isolation on this side, the other on that, without any community between them. Against that view it must be pointed out, that truth is not like stamped coin that is issued ready from the mint and so can be taken up and used. (Hegel Werke 3, 40/Hegel 1977, 22)

This passage echoes the one quoted above on Aristotle's *De Interpretatione*, when Hegel states that conceptual truth requires reflection, i.e. thought thinking about itself, and thought thinking about itself is one and the same as critique and negation of the pretended truth of our assumptions.²³

---

22 I have slightly changed the 1977 translation, using, instead of "Notion(s)", "concept(s)".
23 Nuzzo 2011, 99ff. highlights that conceptual truth implies a challenge to the *linearity* of truth typical of *Vertsandeslogik*. Conceptual truth is not a fixed given but is established through and as a process. The articulation of this process is the task of Hegel's *Vernunftlogik*, and in particular of his *Begriffslogik*. Nuzzo writes that "the speculative form of truth [...] is the structure according to which (i) an advancement is made, and (ii) an ascending oriented movement with a higher and a lower level is established. (iii) Moreover, since that which is overcome in the structure of truth specifies or determines this very truth as the ‚truth of', that which precedes and is overcome (i. e. the false) is still present within truth. Clearly, this structure fundamentally alters the linear and static opposition of truth and falsity defended by the logic of the understanding" (Nuzzo 2011, 101). Speculative truth does not leave anything behind; it is cumulative and inclusive; it is con-

7 Truth-bearers

When we want to find the truth about something we have to reflect on our own assumptions. We cannot simply stick to what we state to be true, rejecting what we think to be false, but we have to take the negation of our original assumption, which we originally held to be false, into careful consideration:

> This truth therefore includes the negative also, what would be called the false, if it could be regarded as something from which one might abstract. The evanescent itself must, on the contrary, be regarded as essential, not in the determination of something fixed, cut off from the true, and left lying who knows where outside it, any more than the true is to be regarded as something on the other side, positive and dead. Appearance is the arising and passing away that does not itself arise and pass away, but is 'in itself,' and constitutes the actuality and the movement of the life of truth. The true is thus a vast Bacchanalian revel. (Hegel Werke 3, 46/Hegel 1977, 27)

– *The negative.* The "false", "apparent", "negative", or "evanescent" is thus essential, according to Hegel. And in so doing he refers to the fact that essences – concepts, universals, forms – are not rigid, but living and dynamic determinations. If we merely said: the false, for example the sentence p "Berlin has been taken up by monsters", is essential, but conceived what is essential in static terms, then it is as if we said: the false, namely that Berlin has been taken up by monsters, is true, and that would be all. That the false is essential for Hegel means rather that the sentence p belongs to a complete and self-critical thought about Berlin, e. g. insofar as we derive from its negation the complete truth about Berlin.

As a matter of fact, it is important to note that Hegel, in the same context of the preface in which he criticises the sentential form and stresses that "truth is not a minted coin", also maintains that

> on this point it may be mentioned that the dialectical process likewise consists of parts or elements which are sentences. The difficulty indicated seems therefore to recur continually, and seems to be a difficulty inherent in the nature of the case. (Hegel Werke 3, 61/Hegel 1977, 40)

Thus Hegel underlines not only the limits but also the necessity of expressing (dynamic and complete) conceptual contents sententially. For Hegel it is rather the case that we use sentences to express both speculative and non-speculative contents. For this reason, in the preface of the *Phenomenology of Spirit* Hegel pos-

---

crete in that it uses the false as the means to acquire determinateness and specification and ultimately completion (*Vollendung*).

tulates the idea of a particular type of sentence, which he calls "speculative".[24] In his later works the Hegelian conception of the speculative sentence develops into the distinction between two kinds of sentences, *Satz* and *Urteil* and, correspondingly, between two kinds of truth.

In sum, all this does not mean that the truth-predicate and the conceptual function we associate with this term do not concern sentences or judgements, for Hegel. It thus does not prove the total incompatibility of Hegel's theory of truth with contemporary philosophical logic, either. "The true is the whole" means rather that sentences have a partial nature, i.e. that sentences as such, in logic, are neither true nor false but need to be *completed* in order to be true. In order to clarify this point, it is useful to consider further Hegel's distinction between *Satz* and *Urteil*.

7.4 *Satz* and *Urteil*

Hegel distinguishes between *Satz* and *Urteil*, and correspondingly between two kinds of truth, *Richtigkeit*, usually translated as "correctness" – but I also use "bare truth" or "simple truth" – and *Wahrheit*, truth in the specific philosophical and conceptual sense.

As we have seen, the corresponding English term for *Satz* is "sentence/proposition" while that for *Urteil* is "judgement". In German, the first term derives from the verb *setzen*, which means positing – *Satz* would be the sentence intended as a simple position in language of something, e.g. a view, a belief; it corresponds to the Greek *thesis*. As for the term *Urteil*, there are controversial views about its etymological derivation.[25] According to one view, whose first proponents were Hölderlin and Hegel, the term comes from the German verb *teilen*, which means "dividing", and the suffix *ur*, which means "original". An *Urteil* is thus an original division. The word was used for the first time in this meaning

---

[24] On the development of Hegel's conception of the speculative sentence see Düsing 1976, in particular 64 ff. and 198 ff.
[25] According to Ritter/Gabriel/Gründer (eds.) 1971 ff., vol. 11, 430 – 461, *Urteil* belongs to juridical terminology, and comes from the juridical old German verb *irtelian* which means *erteilen eines Gersichtsspruchs* (communicating a verdict). The specifically logical use of the term goes back to Wolff's translation of the Latin "*iudicium*" as "*Urteil*" in his German Logic in the sense of "*logos apophantikos*" and is also typical of both Kant and Hegel. In Wolff's German Logic *Satz* is the translation of the Latin *propositio* and has the meaning of "Setzung" position.

by Hölderlin 1795 (Hölderlin 1961, 216 f.).[26] Hegel, who shares Hölderlin's idea, explains in § 166 of the Encyclopaedia that

> the etymological meaning of the judgment (*Urtheil*) in German goes deeper, and expresses the unity of the concept as the first, and its distinction as the original division, which is what the judgment really is. (Hegel Werke 8, 316/Hegel 1991, 244)

For Hegel, in the expression *Urteil* (or according to the old German use *Urtheil*) the strictly conceptual point of view emerges. As we will see better in what follows, for Hegel the conceptual content is unitary but innerly heterogeneous, includes differences and oppositions, and the judgements aiming at expressing the meaning of a concept give voice to this heterogeneity, and more specifically to the "division" internal to every conceptual content.

In the chapter on the judgement in the *Science of Logic* Hegel further specifies the meaning of the expression "*Urteil*" and its difference from "*Satz*":

> We may take this opportunity of remarking, too, that though a proposition [Satz] has a subject and predicate in the grammatical sense, this does not make it a judgment [Urteil]. The latter requires that the predicate be related to the subject as according to the conceptual determinations, that is as a universal to a particular or individual. If a statement about a particular subject only enunciates something individual, then this is a mere proposition. For example, 'Aristotle died at the age of 73, in the fourth year of the 115th Olympiad,' is a mere proposition [Satz], not a judgment [Urteil]. *It would partake of the nature of a judgment only if doubt had been thrown on one of the circumstances, the date of the death, or the age of that philosopher.* (Hegel Werke 6, 305/Hegel 1969, 626)

In a *Urteil* "the predicate relates itself to the subject according to conceptual determinations, that is as a universal to a particular or individual determination". In a *Satz*, by contrast, there is no trace of universality or generality, i.e. of conceptuality strictly speaking, since the predicate expresses a particular determination of an individual such as Aristotle died in the 4th year of the 115th Olympiad. The difference is also explained with the example: "'my friend N. is dead' is a *Satz* – it is a *Urteil* when the question arises as to whether he is really dead or only seemingly dead". So we see that *doubts about the adequate ascription of a predicate to a subject turn the Satz into a Urteil.* "Urteil" is thus in this sense a "Satz" that is located in a dubitative and critical context. Evidently, every sen-

---

[26] See also Henrich 1965–66, 73–96, Düsing 1976, 66 ff. and Lau 2004, 161. For Bachmann, Gruppe and Bolzano Hölderlin's and Hegel's derivation of the meaning of *Urteil* from *Ur-Theilung* is problematic – they all favour the juridical meaning of *urteilen* as derived from *erteilen* (giving a verdict). See Ritter/Gründer/Gabriel (eds.) 1971 ff., vol. 11, 443.

tence can be doubted. Insofar as we locate sentences in a dubitative context, asking the question "are they true?" we find ourselves in the conceptual field.

Notably, Hegel is here circumscribing the kind of sentences he is discussing in his *Science of Logic:* the sentences he is speaking about, he states, are those in which concepts are at stake. They have the following two features: 1. in them the predicate stands for a universal or general property, such as "being good", "being true" and not an individual or particular one, such as "being born in Turin on August 18, 1996". 2. There are doubts about the attribution of the predicate to the subject, e.g. "Juliette was born in Berlin" is a *Urteil* and not a *Satz* insofar as there are doubts as to whether Juliette was really born in Berlin.

The relevance of this distinction is to be referred to the idea of sentences as ways of expressing conceptual contents. A sentence like *Aristotle died in the 4th year of the 115th Olympiad* is not an *Urteil* because it does not express a conceptual content. Conceptual contents are according to 1. universal and, according to 2., controversial, that is, they include different and opposite determinations. Evidently, the second point refers to the sceptical, negative and reflexive nature of conceptual truth, which we have seen considering the meaning of the whole/process claim. *Aristotle died in the 4th year of the 115th Olympiad,* is, strictly speaking, not an expression of a concept. It would become such, however, if we started to ask: "is it true that Aristotle died in the 4th year of the 115th Olympiad?". As a matter of fact, if we ask this question we, according to Hegel, engage ourselves in the process of analysing the "concept" of Aristotle, i.e. Aristotle's properties, the characters of his life and death etc. But individual terms such as "Aristotle" or "this rose" are not the principal research field of logic, in Hegel's view.

7.5 *Richtigkeit* and *Wahrheit*

Hegel's distinction between "two truths" *Richtigkeit* and *Wahrheit* is useful to clarify what is the logically and philosophically relevant form of truth.

I suggest translating *Richtigkeit* using the expression "bare" or "naked truth". In so doing, I follow Hegel's use in the preface to the *Phenomenology of Spirit,* when he calls "bare" or "naked truths" sentences about "how many feet make a furlong and when Caesar was born".[27] A usual, good translation of *Richtigkeit* is also "correctness". *Wahrheit* is truth properly speaking, which

---

[27] "Even bare truths of the kind, say, like those mentioned [about how many feet make a furlong and when Caesar was born], are impossible without the movement of self-consciousness" (Hegel Werke 3, 41/Hegel 1977, 23).

Hegel also calls "philosophical", "concrete", "speculative" or "conceptual truth". In what follows I refer to full, philosophical truth [*Wahrheit*] as W and to bare truth [*Richtigkeit*] as R.

In § 172 of the Encyclopaedia (and in the *Zusatz*) Hegel writes:

> It is one of the most fundamental logical prejudices that qualitative judgements such as: "The rose is red" or "the rose is not red"[28] can contain truth (*Wahrheit*). Correct [*richtig*] they may be, but only in the restricted confines of perception, finite representation, and thinking; this depends on the content which is just as finite, and untrue on its own account. But *the truth rests only on the form, i.e. on the posited concept and the reality that corresponds to it* [my emphasis]; truth of this kind is not present in the qualitative judgement, however. In common life the terms truth and correctness are often treated as synonymous: we speak of the truth of a content, when we are only thinking of its correctness. Correctness, generally speaking, concerns only the formal coincidence between our representation and its content, *whatever the constitution of this content may be* [my emphasis]. Truth, on the contrary, lies in the coincidence of the object with itself, that is, with its concept [...] The subject and predicate of [a sentence such as "the rose is red"] do not stand to each other in the relation of reality and notion. (Hegel Werke 8, 323 f./Hegel 1991, 249 f.)

It is important to note that, even if Hegel distinguishes between the two terms *Urteil* and *Satz*, this does not prevent him from using the two synonymously in other contexts, as he does in this passage, when he speaks of "qualitative judgements" (*Urteile*) as not capable of conveying truth.[29] Both *Urteil* and *Satz*, as we have seen, refer to sentences intended as *lógoi apophantikói*, i.e. linguistic structures that we use to convey truth. The difference is rather between sentences that are able to convey *philosophically significant* truth (W) and those that are not. Sentences such as "this rose is red" do not have conceptual form. They can be correct, i.e. be the bearers of R, which means that they involve a perception or representation, such as our seeing a rose in a vase on the table, and the correspondence of this representation with reality: the rose in the vase on the table. *Whatever the constitution of this content might be* means that they do not tell us anything about the concepts at stake. They only say something fortuitous and contextual of something, something that can also not be the case (in

---

[28] By "the rose is red" Hegel means, evidently, *this* rose is red (and not "*all* roses are red") i.e. the empirical and individual rose. See the following passage, in which Hegel uses, as an example of non-speculative judgement: "this rose is red".

[29] Hegel states that in a qualitative judgement we "immediately" unify subject and predicate (predicate something of something) without reflection, i.e. without questioning the legitimacy of the predication. E.g. we state "my friend N. is dead" and there are no doubts about N's death. Hence what Hegel calls "qualitative judgement [*qualitatives Urteil*]" is a synonym for "*Satz*".

this perspective, we may say that bare truths are contingent truths). Hegel himself explains this point with an example:

> We may add that *the untruth of the immediate judgment lies in the incongruity between its form and content* [my emphasis]. To say "This rose is red" involves (in virtue of the copula "is") the coincidence of subject and predicate. The rose however is a concrete thing, and so is not red only: it has also an odour, a specific form, and many other features not implied in the predicate red. The predicate on its part is an abstract universal, and does not apply to the rose alone. There are other flowers and other objects which are red too. The subject and predicate in the immediate judgment touch, as it were, only in a single point, but do not cover each other. The case is different with the conceptual judgment. When we say "This action is good", we state a conceptual judgment. Here, as we see at once, there is not the loose and external relation between subject and predicate which was typical of the immediate judgment. The predicate in the latter is some abstract quality which may or may not be applied to the subject. In the judgment of the notion the predicate is, as it were, the soul of the subject, by which the subject, as the body of this soul, is characterised through and through. (Hegel Werke 8, 323 f./Hegel 1991, 249 f.)

The sentence (as an immediate judgement: "S is P") manifests an incongruity between form and content. The form "S is P" implies that S and P coincide, i.e. that the literal meaning of the copula is the sign "=", or identity. But in the sentence "the rose is red" S (the rose) and P (red) do not fully coincide (the "is" here has a non literal meaning and expresses accidental predication). The rose is an individual thing, and has also other properties than the one of being red. The predicate is a universal, and can be applied to other things than the rose. The meaning of the concept "rose", i.e. of the predicate "being a rose", is not completely determined by the predicate "being red". In a conceptual judgement, in contrast, there is no incongruence between form and content. If I say "this action is good" as a conceptual sentence/judgement, the predicate completely determines the subject. The predicate is, in this case, not simply a property that may or may not apply to the subject, but the "soul of the object". In it, the copula has the literal meaning of essential predication or identity.[30] In this perspective, it is evident that Russell's critique of Hegel's view on sentences is wrong. Hegel is not "confusing the 'is' of predication, as in 'Socrates is mortal', with the 'is' of identity, as 'Socrates is the philosopher who drank the hemlock'".[31] Rather, he explicitly distinguishes between the two, explaining that the philosophically relevant

---

[30] Düsing 1976, 198 and 199 suggests that Hegel's conception of the speculative sentence refers to idea of specifically "philosophical, essential sentences", and corresponds to Aristotle's conception of substantial sentences, expressing the *tò tì én eînai*.
[31] Russell 2009, 48 and footnote.

sentences are the definitional ones, in which the "is" expresses complete determination.

But why is the relation between S and P in the Hegelian example "this action is good" not "loose and external"? It depends on the predicate at stake in the sentence "this action is good" and its relation to the subject. While "red" can be said of other things than the rose, "good" can be said, strictly speaking, only actions, if by actions we intend, as Hegel seems to do here, morally significant actions. I can say "this person is good", but what I mean is that her actions are good, that what she does is good. Moreover, the very essence of (morally significant) actions is that they are good.

In sum, W is for Hegel a matter of concepts, or conceptual knowledge and not of what Hegel here calls "perception", "representation" (*Vorstellung*) or "finite thinking". In sentential or logical terms, the conceptual point of view is the point of view of the essential and complete determination of the content of a concept. In this case the basic sentential form stands for the definition of a concept: e.g. "justice is the advantage of the stronger", "God is an anthropomorphic being", whereby the copula "is" cannot but be the expression of identity, or full equivalence – in logical terms we would use the double conditional: $\leftrightarrow$. The finite point of view is, by contrast, typical of sentences that merely express properties, and not the essential and complete meaning of something. In the sentence "this rose is red" the copula expresses accidental, and not essential predication, partial, and not complete determination. For Hegel, we can only speak in the former of truth in the philosophical, concrete and speculative sense.[32]

---

[32] For a clarification of the meaning of conceptual truth in Hegel see Koch/Oberauer/Utz (eds.) 2003 and Pinkard 2003, 119–134.

8 The meaning of "true"

8.1 Truth as correspondence

Hegel always defines truth in terms of correspondence. In the passage on Kant's logic at the beginning of the "Subjective Logic" he says, as we have seen in Part 2, that

> The [...] explanation, [according to which truth is] the correspondence of knowledge with its object [has] great, indeed supreme, value. (Hegel Werke 6, 266/Hegel 1969, 593)

In the materials Hegel used for his lectures on logic and metaphysics in Nürnberg truth is defined as "correspondence of the concept with existence" (Hegel Werke 4, 84) meaning that what we think is true if and only if it expresses how things stand (existence); it is said to be "correspondence of the concept with its objectuality" whereby "the sentence is where the presentation of the concept in its objectuality, i.e. the field of truth, begins" (Hegel Werke 4, 105). "Truth is presentation of the concept in its objectivity" means, in other words, that truth is the field in which what we think is objective, i.e. corresponds to how things stand. In the same context, while distinguishing between *certainty* and *rational knowledge*, Hegel remarks that

> The *knowledge of reason* [my emphasis] is not mere subjective *certainty*, but rather also *truth*, because truth is the correspondence or rather the unity of certainty and being or objectuality. (Hegel Werke 4, 123)

Significantly, Hegel identifies here the rational field (the "knowledge of reason") with the realm of truth. Rational knowledge is indeed different from certainty. Certainty is merely subjective, while rational knowledge or "truth" is correspondence between what we think and being, or what really is. In other words, we could say that for Hegel when we are certain about something we only think that what we say expresses how things are, but things could also be different. When we rationally know something, then what we know is necessary and things have to stand just in the way we think they stand.

Hegel thus stresses that truth is correspondence between concept and reality and that "reality is what comes first":

> In every form of knowledge the fundamental element is reality as what comes first and as what is the essence to which the concept must correspond in order to be true. (Hegel Werke 4, 203)

Additionally, he claims that:

> The *truth* of my representations is that they correspond to the constitution and the determinations of the object. (Hegel Werke 4, 213)

And that:

> Certainty as such is not yet *truth*; because truth is the correspondence of my knowledge with the object. (Hegel Werke 4, 291)

Moreover, as we have seen, Hegel points out that the structure of things is sentential and that there is a correspondence between the structure of sentences and the structure of reality:

> All things are sentences – i.e. they are individuals, which are in themselves a universality [...] or they are pure universal, which is individualised. The universality and the individuality is distinguished in them, but is at the same time identical. (Hegel Werke 8, 319/Hegel 1991, 246)

There is no need to go into all of the aspects (the meaning of individuality vs universality, the Hegelian concept of concept, the epistemology of truth) evoked in these passages in greater depth. For now, it is sufficient to stress that in these passages "true" means, classically, "corresponding with reality". These passages correlate with others contained at the end of the "Subjective Logic" in the *Science of Logic*, were the realm of rational thought, i.e. the realm of truth, is presented.

The question is whether Hegel's view about correspondence can be taken in the normal, platitudinous terms implied by standard correspondentism. In the *Science of Logic* and the *Logic* of the *Encyclopaedia* Hegel seems to refer to a peculiar correspondence, the one of "the concept with itself", of "the object with its concept" and "the content with itself". Here he also refers to the already mentioned distinction between W and R.

In the *Encyclopaedia*, in the context of an explanation concerning the meaning of "logic", Hegel writes:

> We must however in the first place understand clearly what we mean by truth [*Wahrheit*]. In common life truth means the correspondence of an object with our representation. We thus presuppose an object to which our representation must conform. In the philosophical sense of the word, on the other hand, truth may be described, in general abstract terms, as the correspondence of a content with itself. This meaning is quite different from the one given above. At the same time the deeper and philosophical meaning of truth can be partially traced even in the ordinary usage of language. Thus we speak of a true friend; by

which we mean a friend whose manner of conduct accords with the notion of friendship. In the same way we speak of a true work of art. Untrue in this sense means the same as bad, or self-discordant. In this sense a bad state is an untrue state; and evil and untruth may be said to consist in the contradiction subsisting between the determination or concept and the existence of the object. Of such a bad object we may form a correct representation, but the import of such representation is inherently false. Of these correctnesses [*Richtigkeiten*], which are at the same time untruths, we may have many in our heads. (Hegel Werke 8, 86/Hegel 1991, 59f.)

By truth we mean the correspondence of our representations with reality; in the philosophical meaning truth is correspondence of a content with itself. When we are talking about representative contents, such as roses in vases on tables, cats on mats etc. truth, here meaning correctness – *Richtigkeit*, consists in the correspondence of what we represent with those represented objects. By contrast, when we are talking about non-representative, conceptual contents such as friendship, the good, justice, truth, being etc., truth as W consists, Hegel says, in the correspondence *of a content with itself*. Hegel thus recalls some examples of the second meaning of "true": we can call a friend a true friend or a work of art a true work of art and this means correspondent to the very idea of friendship or work of art. "Not true" would mean in this sense bad, "inadequate in itself". Accordingly, a bad state is an untrue state and the untruth consists in the contradiction between the concept of the state and the actually existent state.

But if, as Hegel says, correspondence means correspondence of the content with itself, one could object that Hegelian truth is not correspondence traditionally intended as relation between the categorically distinct terms "thought" and "reality", but rather something else, for example coherence. As a matter of fact, this and other similar observations have led interpreters to stress that Hegel did not hold a correspondence conception of truth.[33] More specifically, it has been argued that truth for Hegel is not a property of propositions/sentences but rather of "things". Stern 1993 refers to a distinction, which he traces back to Heidegger, between propositional and "material" truth. What Heidegger calls the theory of "propositional truth" implies that truth is a property of sentences and means "correspondence with the way things are"; what he calls theory of "material truth", by contrast, holds that truth is a property of things and that a thing is true if it corresponds to its essence. In this respect, on Stern's view, Hegel's truth is material truth (a "true friend" is a friend that coincides with the very es-

---

[33] Baldwin 1991, 35–52 writes that Hegel defended an identity theory of truth, according to which a judgement is true if and only if its content is identical to a fact. In any case, an identity theory of truth is not unanimously held as a critique of the correspondentistic point of view, some authors see it as a specific declination of correspondentism (see Glanzberg 2016).

sence of friendship). Baldwin 2004 reacts against Stern, claiming that though the two conceptions are different, they are not incompatible. Rather, according to Baldwin, Hegel grounds propositional truth (*Richtigkeit*) on material truth (*Wahrheit*). In my view W is, from an epistemological point of view, the condition of R, that is, when we have the philosophical, i.e. complete and sceptical truth about something we also have the partial truths about it, and not *vice versa*. However, I stress that W (what both Baldwin and Stern call "material") is to be expressed propositionally, and so cannot but be propositional truth.[34]

In this reading R would be correspondence in normal platitudinous terms, while W would be the specific Hegelian truth that, according to Stern, is not a property of propositions. However, stressing that the Hegelian W is not a property of sentences clashes with both a standardly acceptable meaning of truth, and the Hegelian statements I have presented above, which confirm the standard reading. I suggest instead keeping to the standard, Aristotelian, and also Hegelian insight that truth (in both its form as R and as W) is just the property of sentences; in the second, "material" case truth is the property of specific sentences, sentences that express essences of things or concepts, definitional sentences. Such sentences are true in the classical correspondentistic sense, they are true iff their content corresponds to the way things are. Both a sentence merely expressing R such as "the cat is on the mat" (let us call it *c*) and a conceptual sentence expressing W, such as "the good state is the democratic state" (let us call

---

[34] Schnädelbach 1993 claims that Hegel only superficially accepts the classical conception of correspondence (of knowledge with the object). According to Schnädelbach truth becomes in Hegel "identity of a content with itself". This view is, for him, a sign of Hegel's adherence to a sort of Platonism according to which the correspondence between object and concept, or content and thought, is based on the *metexis* to truth insofar as "everything that is true is true only through truth" exactly like, as Schnädelbach claims, in Plato and Socrates everything that is just and good is so through justice and the good. Even if I share Schnädelbach's general view according to which truth in Hegel is the object of philosophy and that thus giving an account of Hegelian truth implies giving an account of objects, tasks and methods of philosophy, I do not think that the Hegelian approach to truth can be interpreted as Platonic in Schnädelbach's terms. More specifically, I think that the distinction between "the truth" and "everything that is true" is problematic, as it seems to imply a view that endorses a non-propositional, transcendent truth, while for Hegel, as we have seen, both R and W can (and have to) be expressed propositionally. Theunissen 1978, 324–359 interprets Hegel's conception of truth as both *adaequatio intellectus ad rem* and *adaequatio rei ad intellectum*. He stresses that the implicit reference to Kant's Copernican revolution in Hegel's formulation "truth is the correspondence of the object to its concept" (not: our knowledge should correspond to the objects but: the objects should adapt themselves to our knowledge) actually entails a critique of the Kantian standpoint. That "the object corresponds to its concept" is to be intended in non subjectivistic, but rather Platonic terms.

it *d*) are true if and only if they correspond to how things stand. In other words, *d* is true iff it corresponds to "how things stand concerning the meaning of 'state'", i.e. to what a good state is – to the concept of state. And, for Hegel, concepts are real: they belong to reality and are shaped by it. I will come back to this example and the difference between sentences like *c* and *d* in the last section. Now I will consider a question at the core of Hegel's correspondentism: the one concerning the meaning of correspondence itself in Hegel, i.e. of the link between thought and reality.

8.2 The relation between rationality and reality

It is not uncommon to interpret Hegel's view about the link between reality and rationality, paradigmatically expressed by the double sentence "what is rational is real and what is real is rational", as implying a collapse between rationality and reality, along with the view that, for Hegel, reality is constituted by (and not independent from) rationality.[35] Accordingly, Hegel's idealism is often misunderstood as a form of anti-realism.[36] It is not my aim to pursue the question of Hegel's realism, and of the meaning of idealism in Hegel.[37] What is important

---

35 For a clarification of the connection between extra-logical reality and the logical realm in Hegel see Nuzzo 2003, 171–187 (see also Nuzzo 1995, 105–120). Nuzzos analysis is useful to stress how Hegel's logic does not imply the denial of external reality's independent existence, but rather a more subtle view about reality. As Nuzzo shows, Hegel's logic is rather concerned with the dialectics between the conceptual realm and extra-logical reality. Insofar as external, empirical reality is conceived, it loses its extra-logical nature and becomes a formal determination, a determination of thought (Nuzzo 2003, 181ff.). In a way, external reality disappears as soon as it is conceptualised, and re-appears as moment or determination of conceptual thought. On the notion of reality/actuality (*Wirklichkeit*) in Hegel's logic see also Emundts 2018, 387–456 and Ng 2017, 269–290. Emundts (2018, 450f.) highlights the conceptual meaning of *Wirklichkeit* in Hegel and argues for an interpretation of the conceptual realm in terms of inferentialistic holism.
36 d'Agostini 2010, 145f. stresses that Hegel's reception as idealist and anti-realist is deeply rooted in the analytical tradition, and among both Hegel's declared enemies and his declared followers. She writes: "Both determinations may be misleading, since Hegel was not anti-realist and his idealism was philosophical, i.e. transcendental and not metaphysical". Also Rockmore (2010, 158–172) emphasises Hegel's realism, and claims that it is to be read in terms of empirical, and not metaphysical, realism.
37 For an account of the interplay between idealism and realism in some crucial phases of the history of philosophy see Asmuth 2007, 203–221. I share d'Agostini's view (in 2010, 145ff.) about the philosophical nature of "idealism" in Hegel, a view she specifically traces back to Croce's interpretation of Hegel's philosophy (see Croce 2006).

to stress at this point is the Hegelian view on the link between thought and reality (the two terms of the correspondence relation). As we have seen, Hegel suggests that we have truth, i.e. correspondence of thought and reality, not when we are merely certain about something, but rather when we rationally know something.

Thus, provided that "rationality" for Hegel is true thought,[38] we can take the famous double sentence (*Doppelsatz*) as crucial indication about the meaning of truth in Hegel, and in particular about the kind of link between thought and reality that is established by the conceptual function we call "truth".[39]

The famous *Doppelsatz* is introduced in the preface of the *Philosophy of Right*. Its formulation is:

> What is rational is real and what is real is rational. (Hegel Werke 7, 24)

In the introduction to the *Encyclopaedia* (§ 6) Hegel recalls that this sentence "has given rise to expressions of surprise and hostility". He makes clear that:

> As for the term reality (*Wirklichkeit*), it should be evident in what sense I use it, since in a detailed *Logic* I had dealt with reality, accurately distinguishing it [...] [from] other determinations. (Hegel Werke 8, 48/Hegel 1991, 29f.)

---

38 In the Subjective Logic of the *Science of Logic* Hegel calls the adequate concepts (concepts coinciding with reality, i.e. expressing truth) ideas of reason. In so doing, he recalls the Kantian conception of ideas as concepts of reason (*Vernunftbegriffe*) see *Encyclopaedia* § 213ff. (Hegel Werke 8, 367ff./Hegel 1991, 286ff.) Just as for Kant, ideas for Hegel are expression of pure thought, thought thinking about itself, i.e. they are semantic or reflexive concepts (*Reflexionsbegriffe*). Differently from Kant, for Hegel "something possesses truth only insofar as it is idea" and the idea "is the adequate concept, that which is objectively true, or the true as such". Thus for Hegel as well as for Kant the rational realm is the realm of pure thought, thought thinking about itself, but while Kant separates pure thought from truth, Hegel considers it as truth's norm or condition.

39 For an analysis of the *Doppelsatz* see also Stekeler-Weithofer 1992, 35f., where the formal meaning of "reality" (*Wirklichkeit*) in Hegel is highlighted. In other words, the *Doppelsatz* does not mean, for Stekeler-Weithofer, that reality intended as "how things stand" is rational and is to be accepted as it is. Rather, in order to understand the proper meaning of the *Doppelsatz* we must pay attention to the *modal* nature of reality (*Wirklichkeit*) in Hegel, and to the fact that reality, possibility and necessity are, in Hegel, modalities of our thought about things. For Emundts (2018, 453) the *Doppelsatz* expresses that there is a specific difference between reality (*Wirklichkeit*) and existence – something is real/actual (*wirklich*) if it is grasped in its internally differentiated unity, i.e. if it is grasped conceptually.

The "detailed *Logic*" is the *Wissenschaft der Logik*, in which, Hegel stresses, "reality" is dealt with as a determination of thought.[40]

The important point, however, is what Hegel stresses next, namely the question about *the sort of reality we are talking about when we are talking about truth*.

The reality of concepts-ideas

The field of pure thoughts, corresponding to the "what is rational" in the double sentence, is neither too excellent nor impotent, nor "phantasmatic":

> The *reality of the rational* is opposed to both the view that ideas and ideals are nothing but chimeras, and philosophy a mere system of such phantasms and the view that ideas and ideals are something far too excellent to have reality, or something too impotent to procure it for themselves. (Hegel Werke 8, 48/Hegel 1991, 30)

First, concepts are not "impotent" insofar as they are not *only* rational, or something only thought, but also *real*.[41] As Hegel puts it:

> The object of philosophy is the idea: and the idea is not so impotent as merely to have a right or an obligation to exist without actually existing. The object of philosophy is a reality of which those objects, institutions and conditions, are only the superficial outside. (Hegel Werke 8, 49/Hegel 1991, 30)

Concepts (the concept of state, of justice, of right etc.) exist, and their correspondent realised institutions (the real Prussian state Hegel lived in while he wrote his *Philosophy of Right*) have merely contingent and superficial external existence. We could say that the concept of state is the object of a necessary truth (W), while its realised institutions (the Atenian state in the 4th century b.C., or the Prussian state in the 19th century) can be object of contingent truths (R). Notably, the Hegelian conception implies an enlargement of the concept of reality, which is not only taken to include empirical, but also abstract and conceptual facts. Moreover, Hegel is evidently in some way a realist about uni-

---

[40] Thus, it is important to be aware of the fact, stressed by many interpreters, such as Findlay 1981, 132–139, Berto 2005, Chapter VI, d'Agostini 2010, 145 ff., that, from Hegel's point of view, when one deals with reality from a logical point of view she deals with "reality", i.e. the concept of reality. This, it seems, is also what Hegel recalls here.

[41] For a realist interpretation of Hegel, and in particular in terms of realism as "generous noneism" see d'Agostini 2010, 135 ff. For an interpretation of Hegel as an empirical realist see Rockmore 2010, 158 ff.

versals. He does not think, however, that the universals exist independently from their realisations. They exist (imperfectly) instantiated in the particulars.[42]

Second, Hegel stresses that thought/ideas are not "too excellent". This means that they do not concern a transcendent sphere, but rather reality as it is here and now.[43] In the *Philosophy of Right* he highlights, accordingly, that philosophy "is an inquiry into the rational, and therefore the apprehension of the real and present. Hence it cannot be the exposition of a world beyond, which is God knows where" (Hegel Werke 7, 24). Similarly, in the *Encyclopaedia* he writes: "It is very important that philosophy be fully aware of the fact that its own content is *reality*. We call the first awareness of this content *experience*" (Hegel Werke 8, 47/Hegel 1991, 28 f.).[44]

Reality and philosophical empiricism

The second half of the *Doppelsatz* (what is real is rational) is less controversial and hence Hegel does not address it explicitly. "What is real is rational" means, plausibly, that if something is real (also in an experiential meaning of reality as what affects our senses), then it is graspable through reason, it can become object of true thought. "It would be a misunderstanding" Hegel writes in the *Encyclopaedia* (§ 8) "if speculative philosophy did not accept the old sentence, wrongly attributed to Aristotle *nihil fuerit in intellectu quod non fuerit in sensu*, nothing is in thought which was not in experience" (Hegel Werke 8, 51f./Hegel 1991, 32).[45] Likewise, Hegel observes that:

---

[42] Hegel writes "rationality exists in the world [and] the world is rational" Hegel Werke 11, 433 as well as "all things are sentences – i.e. they are individuals, which are in themselves a universality [...] or they are pure universal, which is individualised" (Hegel Werke 8, 318f./Hegel 1991, 246).

[43] Findlay 1955 rightly claims that Hegel's philosophy is completely free from transcendent metaphysics. Hegel is the philosopher of the here and now, which means that he did not want to go beyond the immanent.

[44] See more generally §§ 6 and 7 of the *Encyclopaedia*.

[45] By claiming that philosophy starts with experience, it can be argued that Hegel defended a form of empiricism, which moves along Kantian lines (according to Kant, similarly, "everything (every knowledge) starts with experience"). Rockmore (2010, 170) holds that Hegel was an empirical realist in this sense, which means, according to him, that Hegel rejected metaphysical realism: "Hegel refuses metaphysical realism in limiting knowledge to the science of the experience of consciousness, in short to what is given in conscious experience". While I agree with Rockmore about Hegel defending a form of empiricism, it would be misleading to interpret this as meaning that we cannot know reality as it is in itself, or that, since philosophy deals

> Experience is the real author of growth and advance in philosophy. For, firstly, the empirical sciences do not stop short at the mere observation of the individual features of a phenomenon. By the aid of thought, they are able to meet philosophy with materials prepared for it, in the shape of general uniformities, i.e. laws, and classifications of the phenomena. When this is done, the particular facts which they contain are ready to be received into philosophy. (Hegel Werke 8, 55 f./Hegel 1991, 35 f.)

Hegel often uses the expression "external reality". In the *Lectures on the History of Philosophy* he writes:

> [I]n common life all is real, but there is a difference between the phenomenal world and reality. The real has also an external existence; this displays arbitrariness and contingency, like a tree, a house, a plant are there together in nature [...] in order to know what is, it is necessary to go beyond the surface [...] The temporal and transitory certainly exists, and may cause us trouble enough, but in spite of that it is no veritable reality. (Hegel Werke 19, 111/Hegel 1892 ff., vol. 2, 95 f.)[46]

The "reality" philosophy deals with is not rough reality, but rather the concepts or ideas philosophy draws from scientific and non-scientific views about reality. These views result from experience filtered by both scientific and non-scientific thought and thus prepared for philosophy.

---

with what is given to us in experience, the reality it deals with does not exist in itself (i.e. independently) but rather only for us. Thus following Rockmore I grant that Hegel is an empiricist, but, differently from Rockmore, I claim that for Hegel this means just that through experience we know reality, and that reality is given to us in experience. As a matter of fact, Hegel never denied that reality is there independently of our experiencing and thinking it. Thus I would suggest that Hegel was a speculative empiricist rather than an empirical realist. The term was coined by Hegel himself to stress the speculative nature of Aristotle's empiricism, i.e. the fact that Aristotle was able to extract the concept from (the analysis of) experience. See on Hegel and Aristotle's speculative empiricism also Redding 2017, 165–188. Stekeler-Weithofer 1992, 40 f. stresses a similar point, claiming that Hegel endorsed Kant's transcendentalism, i.e. the view that the task of philosophy is analysing the meaning of the conceptual words at the basis of our reasoning, thinking and acting. Differently from Kant, Hegel was, for Stekeler-Weithofer, a *radical empiricist* insofar as he thought that the conceptual words or categories belong to the field of experience. Horn (1982, 133) claims that Hegel endorses a Leibnizian (rather than Kantian) notion of experience as *grounded phaenomenon*, in which "*necessity is given empirically*".

46 I translate here *wahrhafte Wirklichkeit* as "veritable reality" and not, as in other translations, as "true reality". There is a difference between "wahr" (true) and "wahrhaft" (veritable, that can be said to be true). By sticking literally to the German original, possible misunderstandings about Hegel's supposed lack of awareness about the semantic nature of the predicate "true" are prevented. Hegel's specific idea that only reality moulded by conceptual thought is the content of a true sentence/knowledge also emerges more clearly in the literal translation.

All this plainly shows, though at a preliminary level, that Hegel did not deny reality's independent and external existence, but rather stressed that reality as "what is there" and as what has not already been thought simply needs to be thought. Empirical sciences, as well as our natural logic and metaphysics, order external contingent reality through thought, and in so doing prepare it for specifically philosophical inquiry. The characteristic aspect of philosophical inquiry is also that of the Hegelian *Wissenschaft der Logik,* and consists in analysing and developing the field of pure thought (which Hegel also calls rational, conceptual, speculative thought) i.e. thought thinking about itself.

9 Hegel's concept of truth in contemporary perspective

What I have tried to reconstruct so far is Hegel's general view about the notion of truth. I have focused on the questions "what are truth-bearers, for Hegel?", "what does the term 'true' mean, in Hegel?" and, given Hegel's statements on the correspondentistic nature of truth: "what is Hegel's view on the two elements of the classic correspondence relation, thought and reality?". My aim was to stress those aspects of the concept that might be most interesting for our contemporary perspective.

On this basis, I can locate now Hegel's conception of truth within contemporary debates. In what follows I first distinguish Hegel's theory of truth from the coherentistic and the pragmatistic conceptions, conceptions with which Hegel's view is often associated, though wrongly so, in my view. Second I specify the Aristotelian and Tarskian character of Hegel's conception, showing how the peculiarly Hegelian traits (such as the insight according to which single sentences are partial and thus false, the distinction between *Satz* and *Urteil*, and between R and W) do not imply a departure from both Aristotelian and modern logic. Finally I reconsider the basic features of the Hegelian conception of truth by presenting it as a theory about the meaning and role of truth in logic.

9.1 Coherentism or pragmatism?

As we have seen many authors suggest that, even though Hegel defines truth in terms of correspondence, what he means by "correspondence" is not a relation between being and thought. Puntel interprets Hegel's talk about correspondence in terms of relation between being and being, and sees in it a sign of the total lack of semantics typical of his philosophy, and thus its complete incommensurability with contemporary truth-theories (Puntel 2005, 208–242). Other authors read it in terms of a relation between thought and thought, thus stressing the idealistic and coherentistic meaning of Hegel's view.

I think that both approaches are unconvincing. I have shown that Hegel's statements about truth can be taken literally. More specifically, in my view the core of Hegel's truth theory is classical (Aristotelian) correspondentism. A reading of Hegelian truth in terms of a coherentistic or pragmatistic truth theory, as the dominant ones up to now have tended to be, might be misleading.

– *Hegelian truth cannot be seen as holistic-coherentistic truth.* A holistic-coherentistic view implies that single sentences are inadequate to express conceptual contents. More specifically, as Joachim, a paradigmatic proponent of the holistic coherentist point of view, puts it: "Truth in its essential nature is that systematic coherence which is the character of a significant whole [...] Its parts are through and through in the process and constituted by it [...] The coherence [...] is a form which through and through interpenetrates its materials; and they retain no inner privacy for themselves in independence of the form" (Joachim 1999, 50f.). Typically, as Russell (1906–1907, 28–49) claims against Joachim, if we think that only the whole is true and its parts are false then we have to ask "is the (simple) truth according to which simple truths are false true?". If it is, then it is simply not true that simple sentences are false, and the holist-coherentist theory is thereby refuted. Russell (1906–1907, 78) puts this as follows: "There are in the above theory certain intrinsic difficulties which ought to make us suspicious of the premises from which it follows. The first of these difficulties [...] is that if no partial truth is quite true, it cannot be quite true that no partial truth is quite true".

Hegel's view, as I tried to reconstruct it, is different. As we have seen Hegel claims – similarly to a holistic-coherentist – that truth is the whole, and that simple sentences are inadequate. However, this conception does not imply that propositions/sentences are not truth-bearers, or that the properties of a whole cannot be expressed sententially. Quite the opposite, as we have seen Hegel points out that we have only sentences/propositions to express the whole and that the sentence is thus the locus (bearer) of truth.

Moreover, it is reasonable to admit that Hegel was aware that in logic – that is when we "use" truth – we are not interested in "this rose is red" as such, but in the inferential relation this sentence has with other sentences, such as "there are red roses". Thus truth in logic concerns nets of sentences and their inferential relations with each other. However, this does not mean that the locus of truth is not the sentence/proposition, since we can (and do) express the nature of the same inferential net sententially.

– *Hegelian truth cannot be intended in empiristic-coherentist terms.* A version of anti-realism is a fundamental motivation for endorsing a coherentist position. According to an empiristic-coherentist position, sentences and reality are heterogeneous. For this reason we cannot compare sentences with objects or reality, but only combine sentences with each other.[47] As Hempel 1935, resuming Neurath's position, observes: "each sentence can be combined or compared with each other sentence, but sentences are never compared with 'reality' or

---

47 See the reconstruction of the antirealistic motivations of coherentism in d'Agostini 2011, 55 ff.

'facts'".⁴⁸ Significantly, Hegel did hold that the reality with which sentences might or might not agree is reality as grasped by thought, what we think about reality. And what we think about reality is determined by the impact of reality on us. However, this does not mean, in Hegel, that reality is ontologically dependent on thought or unknowable as it is in itself.

– *Hegelian truth cannot be interpreted in pragmatistic terms.* Though not denying that true knowledge has/can have an impact on action,⁴⁹ Hegel does not define truth in terms of usefulness or success. He cannot therefore be considered a defender of a pragmatistic truth-theory. Surely there are pragmatic components in the first part of Hegel's famous dictum "what is rational is real and what is real is rational" (Hegel Werke 7, 24). As discussed earlier, "what is rational is real" means that the realm of conceptual thought, which is the realm of truth, has the tendency of becoming real, and thereby has an effect on reality. This means that our true scientific knowledge about the world (for example about climate change) contributes to shape the world and is useful to it (e.g. in that we start using alternative energies).

Thus, while Hegel would not contest William James's claim according to which "the possession of true thought means everywhere the possession of invaluable instruments of action", he would deny that "our account of truth is an account of processes [...] having only this quality in common, that they pay" (James 1999, 61). In this sense, in the passages considered above it is evident that Hegel shares the common meaning of truth as correspondence and that he does not define truth in terms of usefulness. He thus would endorse Russell's critique of James according to which "the word 'true' represents for us a different idea from that represented by the phrase 'useful to believe', [...] therefore, the pragmatic definition of truth ignores [...] the meaning commonly given to the word 'true'" (Russell 1999, 75). On the basis of the textual analysis, it clearly emerges that Hegel endorsed the common definition of truth in terms of correspondence, but in an enlarged and differentiated way, which, possibly, also includes pragmatistic aspects.

This can be easily explained by considering, as I have repeatedly emphasised, the typically Aristotelian inspiration behind Hegel's conception of logic.

---

48 "Jede Aussage kann mit jeder anderen kombiniert oder verglichen werden [...] aber Aussagen werden niemals mit 'Realität' mit 'Tatsache' verglichen" (Hempel 1977, 97).
49 See Marcuse 1941, D'Hondt 1968 and Ilting 1973 who all focus on the political (and revolutionary) dimension of Hegel's view and in particular of Hegel's double sentence.

9.2 The Aristotelian core of Hegel's theory of truth

As we have seen, Hegelian correspondentism involves the peculiar distinction between *Satz* and *Urteil* and, correspondingly, between W and R. I have also stressed that these distinctions do not entail a dismissal of the standard correspondentistic view, but merely concern the difference between what I have called "essential" and "accidental", "conceptual" and "non-conceptual" predication.

Truth for Hegel is correspondence between *logos* (thought, rationality) and *on* (being/reality). The distinction between Satz and Urteil, and between R and W, does not imply that truth is something other than correspondence. Let us consider an example:

> c. The cat is on the mat
> d. Democracy is government through public debate

In the first case we have the sentence c., which states something about an empirical thing (the cat). In the second case we have d., which states what a concept (democracy) is. In the first case the sentence is true if things (the cat and the mat) stand as it says; in the second the sentence is true if things (the meaning of the concept of democracy) correspond to what we say about them. In both cases we have truth as correspondence, the difference is the different, i.e. empirical viz. conceptual reality to which the two sentences refer. In this sense for Hegel every sentence reflects the structure of reality. Evidently, Hegel's position implies an enlargement of the concept of reality.

As we have seen Hegel stresses that "all things are sentences – i.e. they are individuals, which are in themselves a universality [...] or they are pure universal, which is individualised" (Hegel Werke 8, 318 f./Hegel 1991, 246). And this is also an insight defended by those philosophers (for example Armstrong 2010, but also Russell and the early Wittgenstein, to whom Armstrong himself traces his own view) who claim that "reality is sentence-like". As Armstrong recalls:

> Wittgenstein said at I.I in his *Tractatus* that the world is the totality of facts, not of things. I think he was here echoing (in a striking way) Russell's idea that the world is a world of facts. I put the same point by saying that the world is a world of states of affairs [...] Interestingly, my own teacher in Sidney, John Anderson, used to argue that reality was 'propositional' and appeared to mean much the same thing as Russell and Wittgenstein. One could say metaphorically that reality was best grasped as sentence-like than list-like. (Armstrong 2010, 34)

In Armstrong's view, that reality is sentence-like means that the world is a world of states of affairs, that is of particulars that instantiate universals. Similarly, in the Frege-Russell account sentences are expressible through predicative functions that stand for universals and are saturated by variables that stand for particulars. When Hegel says that all things are sentences he means that things are universals that have at the same time an individual character, or individuals that have a universal character. And this is what makes them apt to be expressed using sentences.[50]

In this respect, the distinction between R and W corresponds to the distinction between two kinds of truths, in particular, we could say, between contingent and necessary, empirical and conceptual truths. Both, however, reflect the nature of reality.

All this stated, it is possible to further develop the meaning of "correspondence" in Hegel. I have suggested interpreting Hegel's double sentence "what is rational is real and what is real is rational" by focusing on the fact that rationality is for Hegel the realm of true thought. In this light, the *Doppelsatz* can give us essential indications about the meaning of the relation that Hegel sees between thought and reality. The *Doppelsatz* can be read as a biconditional according to which:

(H′) $\forall x \, (RAx \leftrightarrow REx)$ (something is rational if and only if it is real).[51]

---

[50] In this respect, I share the spirit of Berto's analysis in 2005, whose main aim is showing the continuity between Hegel and analytic philosophy by stressing the metaphysical import typical of the first analytical tradition. Berto's main idea – but see also d'Agostini 2008a, 243–270, Varzi 2001, and Tripodi 2015 – is that a certain view about the link between logic and metaphysics is at the very basis of the birth of analytic philosophy, i.e. what he defines as "the idea that linguistic distinctions are informative at the metaphysical level" (Berto 2005, 40). Berto points out that Frege's theory in *Sinn und Bedeutung, Funktion und Begriff, Begriff und Gegenstand* is expression of Frege's "ontology", i.e. of his view that names stand for objects and predicates for concepts, where he understands objects as saturated and concepts as unsaturated entities. Differently from Berto, I stress the classical correspondentistic meaning of Hegel's notion of truth.

[51] See d'Agostini 2010, 135 ff. For d'Agostini (2010, 136) the *Doppelsatz* presents Hegel's answer to the meta-metaphysical debate in analytical philosophy about the meaning of the term "existence" and gives a "restriction rule for the use of the predicate of existence", i.e. it tells us when we can reasonably assign the predicate "real" to something. We can call something "real" when it is rational, that is, for d'Agostini, when it is located within our system of knowledge. I substantially share this interpretation, and would add that "rational thought" for Hegel is "true thought", whereby Hegel spells out the meaning of "true", as we have seen, in terms of correspondence.

I have highlighted that Hegel identifies rationality with the realm of true thought. From this point of view, an interpretation of the *Doppelsatz* in alethic terms becomes plausible:

(H″) $\forall x\ (Tx \leftrightarrow REx)$

which means that something is true if, and only if, it is real. In this light the view parallels Aristotle's conception of truth in the fourth book of the *Metaphysics*. I call Aristotle's thesis A:

(A) To say of something that is that it is, and of something that is not that it is not, is the true, to say of something that is that it is not, and of something that is not that it is, is the false (Aristotle, *Metaphysics* IV, 7, 1011b 26 f. – translation from Barnes 1984).

What is now known as the T-schema, namely:

The sentence "p" is true if, and only if, p

aims, according to Tarski, at grasping Aristotle's thesis (A) in a precise, modern philosophical terminology.[52] According to Tarski the T-schema grasps the behaviour of the truth predicate in our language. In debates on truth the schema has been interpreted in both realistic and antirealistic or deflationary terms. What is interesting now is that H″ does figure as a (minimally) realistic interpretation of the T-schema.[53]

The side from left to right of the biconditional in H″ means that if we think truthfully, then what we are thinking about is real, and to be real means, as we have seen, to be the case, existing. A and H″ have an identical core, namely the idea that to think or speak truthfully means that "we say what is". Moreover, the Hegelian view entails, as we have seen, an enlargement of the correspondentistic

---

[52] Tarski 1999, 118 ff. formulates the schema as: "X is true if, and only if, p" whereby "X" stands for the name of the sentence and "p" for the sentence.

[53] The view known as alethic realism, as it is developed by Alston in 1996, grasps this minimal Hegelian realism. Alethic realism is the view according to which "a statement (sentence, proposition belief …) is true if and only if what the statement says to be the case actually is the case". See Alston 1996, 5. Alston also suggests that a definition of alethic realism in terms of truthmakers-theory is in order, i.e. in terms of the theory according to which "a sentence is true if and only if there is something that makes it true". What is peculiar about both alethic realism and truthmakers-theory is that by endorsing them we are not committed to a view about the nature of reality or of the things that make the sentence true – this commitment would be metaphysical realism. The same can be stated with respect to H″.

conception, which is well expressed in the side from right to left of the biconditional.

The side from right to left in H': that if something is real then it is rational means that what is significantly (logically) real, is not the object of *any* kind of thought, but only of a *philosophically relevant* one, and this means, as we have seen: a sceptical (result of a process of reflexion and self-critique) and complete (includes everything there is to say about something) thought. Only then we can talk about truth.

In other words, if it is the case that the accused person is a murderer, and we say that the accused person is a murderer, we tell the truth. However, in light of Hegel's view of rationality, we are able to find the philosophically relevant truth about the accused person only if we systematically question our assumptions about her, and know everything there is to know about her, such as her past, present and future actions etc., that is: if we have the whole set of true sentences about her and this set continuously undergoes a process of critique and revision. Only then we are able to Aristotelianly "say what is", i.e. to tell the truth.

Now one could object that the view about the necessity both of subjecting our assumptions to continuous critique and of knowing everything there is to know about something are epistemic conditions in order to find the truth, and do not concern the *meaning* of truth. Hegel claims, as we have seen, that truth is the result of a complex development brought about by self-consciousness. However, this does not mean that Hegel's reflections are not relevant from a logical perspective. What is interesting about the Hegelian perspective is that the question about the epistemic conditions of truth has a logical relevance, i.e. is a fundamental aspect we should take into account when we try to define the logical behaviour and the nature of truth.[54]

9.3 Hegel and truth in logic

In the previous pages I have stressed the Aristotelian meaning of Hegel's view on truth. For both thinkers the thesis

---

[54] As we will see, Hegel says that scepticism is the *logical* moment of dialectics, by which he means that it concerns the meaning of negation and double negation in dialectics – see here below Part IV. Hence scepticism, though defining an epistemological position, has a logical meaning, and this is the aspect in which Hegel is most interested.

(A) To say of something which is that it is and of something which is not that it is not is the true, to say of something which is that it is not, and of something which is not that it is, is the false (Aristotle, *Metaphysics* IV, 7, 1011b 26f. – translation from Barnes 1984).

holds. H″ is the Hegelian version of A.

Aristotle's position assumes in Hegel a particular significance, which comes into view in light of the distinction between *Verstandeslogik* and *Vernunftlogik*. Hegel's theory of truth as correspondence is revised and reconsidered within the specifically Hegelian speculative logical approach.

In my account, Hegel's approach does not introduce changes concerning the truth-bearers (sentences are for Hegel truth-bearers). Rather, it concerns the nature of the sentences that bear W. These sentences are for Hegel expressions of conceptual contents. In other words, Hegel specifies that his logic (as *Vernunftlogik*) does not deal with bare (i.e. contingent or empirical) truths, such as "this rose is red", "the cat is on the mat", "Aristotle died at the age of 73, in the fourth year of the 115th Olympiad", "My friend N. is dead". Rather, it concerns sentences that are: universal, by which Hegel means that they are:

controversial (in the sense of affirming a possibility or impossibility, e.g. to use Hegel's example, "my friend N. is dead", where I do not know if he is actually dead);

definitional or *essential* (in the sense of affirming what something essentially, definitionally is: "God is an anthropomorphic being"; and of determining what belongs essentially, and not accidentally to something: "this action is good").

These are, we could say, the sorts of sentences that justify the logical and philosophical enterprise in general, not only in Hegel. They are sentences which, in other words, call for further reflection, clarification, reasoning, justification, or inferential connections with other sentences.

From this point of view it is evident that the truth bearer, for Hegel, is the sentence insofar as it is located within a logically relevant situation, namely a dubitative, critical and universal context.

The theses "the true is the whole" and "the true is the process" do not intend to diminish these preliminary assumptions. Rather, they simply underline a thesis that is well known to the logical and philosophical tradition, namely that *episteme* – i.e. science, logic and philosophy – deals with universals, and involves generalisation.[55]

---

[55] See Aristotle in the *Nicomachean Ethics*: "science is a judgement about things that are universal and necessary [...] [science concerns] those things that are demonstrable (since science involves reasoning)" (VI, 6, 1140a 34f. – translation from Barnes 1984).

Summary

In the third part I address Hegel's notion of truth by focusing on two questions at the core of every truth-theory in contemporary philosophical logic: "What are the truth-bearers?" and "What is the meaning of 'true'?".

In the literature there is still disagreement about how Hegel deals with both questions. The most controversial point concerning the first is that Hegel famously declared the sentential form inadequate to express conceptual, speculative truth. The claim seems to imply a major incompatibility between Hegel's position and the standard logical one, according to which sentences are the basic unities that can be true or false, and the actual bearers of truth. The second problem concerns the second question: Hegel defines truth in terms of *correspondence*, but what he means by correspondence seems to be different from the standard correspondentistic position, according to which a sentence (proposition, or belief) is true if, and only if, it corresponds to reality. In other words, Hegel does not only refers to the correspondence of the concept with the object (or reality, or the world), but also to what, for him, is the genuine correspondence relation, namely the one "of the object with its concept", and of "the concept with itself". These claims are often taken to imply a total lack of semantics, on Hegel's part, and as signs of the fundamental incompatibility between Hegel's logic and current debates on truth theories.

In this part I also examine Hegel's theses on judgements and their link to truth, as well as on the meaning of "true". I show that both Hegel's critique of the sentential form, and his definition of truth, imply neither a dismissal of the classical Aristotelian view of the sentence as *logos apophantikos* nor a rejection of the classical conception of truth as correspondence. On this basis, Hegel's view on truth can genuinely dialogue with contemporary truth-theories, and an assessment of its originality within debates in philosophical logic becomes possible.

In **Chapter 7** I first consider Hegel's claim in the *Logic* of the *Encyclopaedia* according to which, in logic, judgements cannot be taken in subjective or psychological terms, but instead have the meaning of *structures aiming at expressing the nature of reality*. "All things are judgements", Hegel states, and specifies that judgements, as well as reality, are structured in terms of universals (the predicates in the sentences) united with particulars (the subjects). For this reason, Hegel also states that "the judgement is truth" and that "judgements are where truth begins". All this clearly shows that Hegel does not discuss the basic Aristotelian insight according to which judgements (or sentences) are the linguistic unities that can be true or false. Second I examine Hegel's critique of the sentential form in the preface to the *Phenomenology of Spirit*. Here Hegel famously claims that the *form* and *element* in which truth can exist is only its scientific system or, which is the same, the concept, and that the sentence/judgement is inadequate to express the conceptual nature of truth. I explain that Hegel here does not want to question the idea according to which "true" is a property of sentences/judgements. What Hegel is talking about are rather the epistemic conditions of truth. What he wants to highlight is that truth is a property of scientific knowledge. When we are merely certain that p, p could also be false. When we scientifically know p, p cannot but be true. In this sense, Hegel follows the Aristotelian view according to which *episteme* (scientific, universal knowledge) is the very condition of speaking and thinking truthfully. As it is already clear in the preface to the *Phenomenology of Spirit*, scientific knowledge for Hegel is both complete (it implies knowing everything there is to know about something) and sceptical (it involves continuous self-critique). Third, I present Hegel's distinction between *Urteil* (usually translated as "judgement") and *Satz* (usually translated with "proposition"), *Richtigkeit* (correctness, or bare, simple truth) and *Wahrheit* (complete, speculative truth). *Urteile* are bearers of genuine (scientific, conceptual) truth, while *Sätze* are bearers of mere correctness (naked/bare truth). These distinctions show that Hegel's critique of the propositional form cannot be taken as rejection of the view according to which truth is, strictly speaking, the property of sentences/judgements.

In **Chapter 8** I examine Hegel's definitions of "true". Hegel states that the classical definition of truth as correspondence of our knowledge with the object is "of supreme value". At the same time he specifies that *Wahrheit*, as opposed to *Richtigkeit*, is the correspondence of the object with its concept, or of the concept with itself. These claims are at the core of the interpretations of Hegel as coherentist, or pragmatist, or only partly correspondentist. Some interpreters claim that for Hegel truth is not the property of propositions but rather of things. Others lament, on this basis, the complete lack of semantic awareness on Hegel's part. I hold in contrast that Hegel's defence of correspondence should be taken literally, truth for Hegel *is* correspondence in the classical meaning. The difference between the standard and the Hegelian conception is rather that, according to Hegel, the bearers of truth (*Wahrheit*) are conceptual sentences, such as d.: "democracy is government through public debate", and not normal ones, such as c.: "the cat is on the mat". In both cases the sentences are true if and only if they correspond to how things stand (or if there is something that makes them true). But while in c. the reality we are talking about is the common sense, empirical and contingent one, in d. we are talking about concepts. All this means that the difference between a standard and a Hegelian version of correspondentism rather concerns the *kind of reality* conceptual sentences refer to.

I conclude the chapter by analysing the question about the meaning of reality, and more specifically of the link between thought and reality, in Hegel.

In **Chapter 9** I consider all the principal traits of Hegel's theory of truth, confronting them with some conceptions of truth canonical in philosophical logic. I first argue that Hegel's theory cannot be understood in terms of coherentism (neither holistic nor empiristic) nor in terms of pragmatism, two theories with which it is often associated, though wrongly, in my view. By contrast I emphasise instead the Aristotelian core of Hegel's conception. More specifically, my proposal is to read Hegel's *Doppelsatz* ("what is rational is real and what is real is rational") in the sense of a minimal-realist version of Tarski's Truth-Schema, which Tarski introduced in 1944 with the explicit aim of making Aristotle's account of truth more precise. Aristotle's definition of truth (I call it A) is:

> A: "to say of something that is that it is and of something that is not that it is not, is the truth; to say of something that is not that it is, and of something that is that it is not, is the false".

Tarski's schema (T) states:

> T: the sentence "p" ("the sun shines today") is true if, and only if, p (if the sun shines today).

Hegel's *Doppelsatz* (H′) states:

> H′: "a thought/sentence/belief is rational if, and only if, it is real".

I claim that rational thought, for Hegel, coincides with *true thought*, and that reality, for Hegel, is reality grasped by thought. Hence the *Doppelsatz* becomes:

> H″: "a thought/sentence/belief is true if, and only if, it expresses how things stand (reality)".

Hence I argue that H″ involves A and a realist reading of T. Also for Hegel it is the case that to tell the truth means saying "how things stand", and saying "what is". The speculative conception does not intend to question this basic assumption, it rather specifies what kind of thought is necessary in order to genuinely "say what is", and to express what the nature of reality is: conceptual, i.e. complete and at the same time sceptical thought.

I conclude by stressing that Hegel's concept of truth coincides with the concept of truth presupposed in the same logical inquiry. That the bearers of truth, for Hegel, have a conceptual nature means that they are both universal and controversial. Universality and controversial nature are the aspects that call for justification, and hence require inference and the unfolding of the logical inquiry (as analysis of what follows from what).

IV **Validity**

Dialectic in relation to the [logic of the] understanding appears as inconsequence in relation to consequence. (Hegel 1992, 13)

IV Validity

The notion of validity (or logical consequence),[1] is the core notion of logic. Addressing Hegel's view on validity is thus crucial if one wants to assess the actuality of his idea of logic. And yet, the emergence of validity as a technical subject of study is a relatively recent product. The correspondent German term is "Folgerichtigkeit" or "Folgerung", which does not figure in the Hegelian terminology in a technical connotation. It corresponds to the Greek *akoloúthesis*, a word used by Aristotle (though not with a technical meaning), and translated by Boethius as *consequentia*. The term began to assume a technical meaning, appearing as a title of treatises, in Medieval Philosophy. John Buridan's *Treatise on Consequences* is, in this respect, a fundamental reference point. It both presents one of the most sophisticated theories of logical consequence in medieval times and explains all the basic concepts connected with it (truth, supposition, ampliation among others).[2] The concept became the focus of discussions in philosophical logic starting from the first half of the 20th century with the work of Alfred Tarski, Gottlob Frege and Rudolf Carnap, and is still at the core of contemporary debates.[3]

Is it possible to assess Hegel's contribution to these debates? Actually, it is not unusual to complain about the inability of Hegel's logic "to determine the fundamental laws of inferences governing all propositions, whatever their content".[4] Hegel himself seems to reinforce the complaint when he writes, in the *Lectures on Logic and Metaphysics*, that:

> [D]ialectic in relation to the [logic of the] understanding [*Verstandeslogik*] appears as inconsequence in relation to consequence. (Hegel 1992, 13)

---

[1] For Asmus/Restall (2012, 11) "the study of consequence and the study of validity are the same".
[2] As it is already made clear in Buridan's *Treatise*, the theory of the syllogism does not coincide with the theory of logical consequence, which is more general. The theory of validity is the general inquiry into the relation between premises and conclusions in arguments, the distinction between valid and invalid ones, and the individuation of criteria in order to perform the distinction. The chapters on syllogisms in Buridan's *Treatise* apply the discussion of validity to arguments having a special form, syllogistic arguments with two premises and one conclusion.
[3] See Ritter/Gründer/Gabriel (eds.) 1971 ff., vol. II., 960–962.
[4] See Beiser 2005, 161. Redding 2014 shows that these views about Hegel's logic are reductive. In 2014, 281–301 he reconstructs Hegel's discussion of the four syllogistic figures in the *Subjective Logic*, showing that Hegel implicitly gives voice to two notions of validity. The first, traditional one (which Redding calls strong) is modelled on the predicative relation of inherence; the second, modern one (called weak) expresses the relation of subsumption. In what follows, I focus on Hegel's history and theory of the dialectical method, which, in my view, corresponds to Hegel's general notion of validity.

As a matter of fact, logic as traditionally intended (Aristotle's logic is, for Hegel, a logic of the understanding – *Verstandeslogik*) fixes inferential forms as valid and distinguishes them from invalid ones. Dialectical logic, by contrast, entails a genuine critique of logical laws and inferential forms. Thus, if the criteria of validity are exhausted by the traditional logical ones, dialectical logic, which questions classical inferential forms, can be seen as inconsequence.

However, we will see that Hegel's talk about dialectical "inconsequence" does not entail a challenge to validity as intended in contemporary terms. The most common notion of validity in contemporary logic is so called "semantic validity". According to it an inference is valid if and only if it draws true conclusions from true premises. This requisite is also called "truth preservation". I will mention the problems and implications of this notion of validity later. What is worth noting now is that such a conception is prefigured by the traditional Aristotelian conception of valid inference (deduction/*syllogismos*) as

> [S]peech (*logos*) in which, certain things having been supposed, something different from those supposed results of necessity because of their being so. (Aristotle, *Prior Analytics*, I. 2, 24b18–20 – translation from Barnes 1984)[5]

Notably, this Aristotelian conception is not questioned by dialectical logic, despite Hegel's critique of *Verstandeslogik*. In dialectical logic it is the case that "having supposed certain things, others follow with necessity because of their being so". In other words, though involving a reversal of intellectual reasoning, dialectical reasoning is necessary, that is, deductive reasoning. Moreover, Hegel repeatedly states that dialectic has a scientific nature, and should be developed methodically.

The whole problem of assessing Hegel's possible contribution both to the history of and to our current comprehension of the concept of validity thus lies in understanding the exact meaning of dialectical inferences, and in distinguishing them from other kinds of inferences. In order to understand Hegel's notion of validity, in other words, we have to reflect on Hegel's notion of dialectic.[6]

---

5 Mignucci 1995, 36 f. observes that, even if Aristotle did not use the word "logic", he is the philosopher who founded logic as a discipline, since he fixed the notion of valid inference, distinguishing between valid and invalid inferences.

6 In 202+, Chapters 5–6 Dutilh Novaes examines the roots of the notion of logical consequence (and more specifically deduction) in dialectics, which she interprets in dialogical terms. Like Dutilh Novaes, I am interested in highlighting the connection between the dialectical and our current notion of valid inferences. However, I also admit other two notions of "dialectics" beside the one of "the art of dialogue", and specifically: the meaning of dialectics as the logic of contradiction and its meaning as the movement of pure (self-reflexive) thought.

In what follows I address this problem by focusing on Hegel's discussion of dialectic in his *Lectures on the History of Philosophy*. In particular, I claim that the *Lectures on Ancient Philosophy* are a fundamental reference point for assessing the strictly logical relevance of Hegel's thought. Classically, the authors (like Michelet 1888, Gadamer 1976, Düsing 1976, Baum 1988 among others) who underline the significance of the *Lectures* for clarifying the formation of the dialectical method do not engage with philosophical logic.[7] In turn, the authors who read dialectics from a logical point of view (such as Apostel 1979, Kosok 1979, Marconi 1979a, Priest 1989, Berto 2005 among others) do not consider Hegel's *Lectures on the History of Philosophy*.[8] Not only that. While Hegel's published writings contain few definitions of the term "dialectic", which are often very dense and obscure, Hegel's observations in the *Lectures on the History of Philosophy* are exemplarily clear. Hence I first (10.) will consider Hegel's history of dialectic from Zeno to Kant, then (11.) I will summarize Hegel's own account of dialectical validity in the *Vorbegriff* to the *Logic* of the *Encyclopaedia*. Finally (12.) I will directly address the highly controversial question "what is Hegel's dialectic?" assessing the relevance of Hegel's account both within the history of logic and in contemporary debates on validity in philosophical logic.

---

[7] On the importance of ancient dialectics for the development of Hegel's own theory of dialectics see Gadamer 1976, Düsing 1983, Riedel 1990, 13–41, Pöggeler 1990, 42–64. For a complete bibliography on "Hegel and ancient dialectics" see Wasek 1990, 275–283.

[8] Recent exceptions are Butler 2012, d'Agostini 2011b, 121–140 and Redding 2014, 281–301. However, these authors do not develop a systematic analysis of the logical importance of Hegel's *Lectures on the History of Philosophy*, such as the one I am trying to unfold here. Butler's main aim is assessing the logical meaning of Hegel's dialectic, and he does consider the Hegelian *Lectures*. However, he only focuses on the Hegelian reading of Pythagorean thought. d'Agostini assesses the meaning of Hegel's reading of Megarian paradoxes for contemporary debates on truth and paradoxes. Redding is interested in integrating Hegel in the canon of the history of logic, and hints at the importance of his interpretation of Stoic logic in particular.

10 Dialectic from Zeno to Kant

Hegel applied dialectic everywhere in his published works, but he did not write a monographic study on the subject. What is more, in his published writings there are only few definitions of the term. This is, possibly, one of the reasons why dialectic has been the core of endless discussions in the whole history of its reception, ever since the last years of Hegel's academic activity in Berlin. A typical misunderstanding is exemplified by the anecdote of a dialogue between Karl Ludwig Michelet, one of Hegel's students who understood his thought most comprehensively, and his friend Tollin. Michelet observed: "In Hegel one can find the most severe monotheism unified with pantheism, idealism fused in one with materialism". Tollin reacted: "Ah, yes, he turns the coat as soon as the wind turns". "Oh! No!" replied Michelet: "He rather has one coat for every wind!" (Nicolin 1971, 230–231). Tollin gives voice to a typical prejudice against dialectic, identifying it with trivialism (the view that everything is true), or even with opportunism. Goethe himself, who often claimed to be an admirer of Hegel's writings, also admitted that he did not fully understand the meaning of dialectic. Eckermann reports that once, during a conversation, Goethe asked Hegel "what is dialectic?", and Hegel answered: "it is basically nothing other than the spirit of contradiction, which is of fundamental importance in order to distinguish truth from falsity" (Eckermann 1987, 622–623). Both Tollin's reaction and Goethe's question are a sign of the difficulties interpreters often encounter in assessing the meaning of Hegel's philosophical method, while Hegel's answer shows, at a preliminary level, that dialectic must be put in the service of (a seemingly classically understood) truth.

In this panorama, the materials gathered in the *Lectures on the History of Philosophy* are notable, since they entail a complete analysis of the concept and a clear explication of the differences between dialectical and non-dialectical, valid and non-valid inferences.

In the *Lectures on the History of Ancient Philosophy*[9] Hegel distinguishes between many forms of dialectic developed by different authors (Zeno, Heraclitus,

---

[9] The logical relevance of Hegel's *Lectures on the History of Philosophy* also consists in bringing light on a question that Kneale and Kneale in *The Development of Logic* describe as "mysterious": "[In Plato, dialectic is] the hypothetical method of refutation together with some mysterious positive addition" Kneale/Kneale 2008, 10. In this respect the Hegelian account is fundamental, as it allows us to show that the "positive addition" involved in not only Platonic, but every "good" dialectic, including Hegel's own, is not mysterious at all. Priest and Routley also underline that what the Kneales call "mysterious" is not difficult to understand, Priest/

Plato among others). According to Hegel, all these forms "interest philosophy now" (Hegel Werke 18, 275/Hegel 1892 ff., vol. I., 240).

In Hegel's account it is possible to recognise the three fundamental meanings with which dialectic is standardly associated: first, dialectic is intended as the "movement" or "semantic behaviour" or "logical destiny" of concepts, especially higher order concepts: the forms of thought (categories) that are expression of the forms of reality, and that are words deposited in our natural language and reasoning. This meaning is especially stressed by Gadamer – we can call it the conceptual meaning.[10] Second "dialectic" refers to discussive or dialogical confrontations. This can be said the "discussive notion"[11]. Third, dialectic is conceived as the logic of contradictions, and hence as a logic that admits of some true contradictions. I call this the logical meaning.[12] In Hegel's reconstruction the three meanings are intertwined and variously operate in the interpretation of the different authors.

10.1 Zeno, Sophists and Heraclitus

For Hegel dialectic begins with Eleatic philosophy: "We here find the beginning of dialectic, i.e. the pure movement of thought in concepts".[13] According to Hegel, the philosopher who first introduced dialectic as systematic method of inquiry is Zeno.[14]

Zeno practiced dialectic for the first time as "movement of the concept in itself" (Hegel Werke 18, 295/Hegel 1892 ff., vol. I., 261). While Parmenides and the Eleatics simply stated the truth of the one, negating the many,

> [W]ith Zeno, on the contrary, we certainly see just such an assertion of the one and removal of what contradicts it, but we also see that this assertion is not made the starting point; he rather starts with what is commonly established as existent, showing how it nullifies itself. (Hegel Werke 18, 295/Hegel 1892 ff., vol. I., 261)

Routley 1984, 86 footnote 21. However, as I will show in the following, Priest's and Routley's proposal to "solve the mystery" substantially differs from the Hegelian one.
10 See Gadamer 1976.
11 This notion is recognisable in the works of Hintikka 2007 and van Eemeren/Grootendorst 2004.
12 This notion is stressed in paraconsistent interpretations such as Priest/Routley 1984.
13 Hegel Werke 18, 275/Hegel 1892 ff., vol. I., 240. On Hegel's interpretation of Eleatic philosophy see Berti 1990, 65–83 and Bubner 1990, 84–97. On Hegel and Heraclitus see among others Boeder 1990, 98–108. On Hegel's interpretation of Sophistic dialectic see Held 1990, 129–152.
14 See Berti 2015, 15–40.

Zeno's aim is to defend Parmenides' view according to which only the immutable one is real, and motion and change are not real. Both Parmenides and Zeno, as Hegel says, "assert the one and remove what contradicts it". Zeno, however, does not begin his argument with this assertion, but rather with its negation, that is with the thesis defended by Parmenides' adversaries (i.e. that the many, or sensible being, space, time, motion etc. exist), or, as Hegel says here, "what is normally held to exist". Zeno thus starts by assuming, for example, that the many (or change or motion) exists, and then shows how such a view "nullifies itself". He achieves this nullifying result by *analysing* the concept in question, examining effective or imaginary cases (the race between Achilles and the tortoise, the Arrow Paradox, the Stadium among others), and then showing how, from this analysis, irreducible contradictions arise. In so doing, he manages to refute the view of Parmenides' adversaries.

In this respect, Hegel also distinguishes between *external dialectics* and *immanent (genuine) dialectics*. The former is the one performed by the sophists: a manner of "reasoning from external grounds", of confounding concepts, granting for instance that "in the right there is what is not right, and in the false the true (Hegel Werke 18, 303/Hegel 1892 ff., vol. I., 265). The latter is

> [T]he immanent contemplation of the object; it is taken for itself, without previous hypothesis, idea or obligation, not under any outward conditions, laws or causes; we have to put ourselves right into the thing, to consider the object in itself, and to take it in the determinations which it has. In this consideration, the object manifests itself as having opposite determinations, and thus breaks up [*er hebt sich selbst auf*]. (Hegel Werke 18, 303/Hegel 1892 ff., vol. I., 265)

What Hegel calls "object" is thus the conceptual content that is to be analysed (the concept of change, motion, the one, the many etc.). While Zeno is only interested in the question: "What is X (motion, change etc.)?", the sophists (as we will see later) were moved by other purposes, such as personal interests or contextual reasons. As a matter of fact, the sophists were, in Greece, similar to lawyers, they were paid in order to defend a particular view (the one of their "clients"). Hence while Zeno "found" contradictions arising within the analysis of the concept, the sophists "produced" them *ad hoc*. That Zeno was interested *only* in the concept also explains why Hegel calls dialectic "movement of pure concepts" or of "pure thoughts". "Pure" refers to the fact that in dialectics we are *only* and *exclusively* interested in the concept, in finding its true meaning, and in nothing else.

Zeno's analysis of motion is an example of this immanent, genuine dialectics. According to Zeno, however, the dialectical movement ends with the nullification of the analysed concept, "the affirmative in it does not yet appear"

(Hegel Werke 18, 303/Hegel 1892ff., vol. I., 265). By contrast, the specifically Hegelian dialectic implies that the destructive result of the conceptual analysis positively tells us something about the true nature of the conceptual content at stake. A further distinction can thus be fixed between two kinds of immanent or internal dialectics: *internal dialectic with negative result* and *internal dialectic with positive result*. The first is the one practiced by Zeno, the latter is the Hegelian one, prefigured first by Heraclitus (and then by Plato).

In discussing the philosophy of Heraclitus, Hegel famously observes that "here we see land; there is no proposition of Heraclitus that I have not adopted in my logic" (Hegel Werke 18, 320/Hegel 1892ff., vol. I., 279). He sees in Heraclitus' philosophy the positive dialectics that he found lacking in Zeno. However, in the chapter about Heraclitus in the *Lectures on the History of Philosophy* the focus is not primarily on dialectics, but rather on a particular case of dialectical development, namely the dialectic of being and nothing, and on the concept of "infinite". Whereas Parmenides observed that "only being is and non-being is not [...] you can neither reach nor know nor express non-being", Heraclitus recognises that being is as much as non-being is and that "truth is the unity of the opposites". This concept of truth as unity of opposites is the reason why Heraclitus assumes such a fundamental role in Hegel's thought.

In this section, however, I will limit myself to clarifying the meaning of dialectic, and the difference between Zeno's and Heraclitus' attitude towards contradictions.

Zeno practices for the very first time dialectics as the method of detecting contradictions within pure concepts, but he considers what is contradictory as false. Thus Hegel states that:

> Zeno [...] shows the opposition within [the concept of] movement [...] he expresses the infinite, but on its negative side only, because he takes its contradiction as being the untrue. (Hegel Werke 18, 325/Hegel 1892ff., vol 1, 282)

While

> in Heraclitus we see the infinite as such, or the expression of its concept, essence: the infinite [...] is the unity of the opposites, in particular of the pure opposites: being and not-being. (Hegel Werke 18, 325/Hegel 1892ff., vol 1, 282)

Interestingly, Hegel's talk about "the infinite" and his distinction between a genuine and a bad infinite parallels his view of dialectics, particularly his distinction between genuine and bad dialectics.[15]

For Hegel, the sophists discovered the destructive power of conceptual thought:

> [The Sophists found] the concept [...] as the absolute power in front of which everything vanishes; and thereby all things, all existence, everything held to be secure, is now made fleeting. (Hegel Werke 18, 406/Hegel 1892ff., vol. I., 352)

In Hegel's view the Sophists discovered that conceptual analysis and conceptual thought ("the concept", "*der Begriff*") is nullifying and sceptical. Very simply, this means that when we analyse a concept, whether it be empirical or non-empirical, e.g. the concept of honey, we start proposing possible other predicates that make the meaning of the concepts at stake explicit, "is sweet", "is liquid" etc.. If we go on thinking, however, other aspects emerge, which possibly deny the first ones, e.g. "is bitter if compared to sugar", "thick if compared to water" etc. The sophists discovered, according to Hegel, the destructive and sceptical power of conceptual thought.

10.2 Plato

Plato was well aware of the nullifying power of conceptual analysis. However, his perspective was a radically new one. Hegel distinguishes between Plato's genuine dialectic and the empty dialectic of the sophists.[16]

> Plato's inquiry is focused on pure thoughts. Dialectics means considering pure thoughts in themselves [...] such pure thoughts are: being and non-being, the one and the many, the infinite (the unlimited) and the limited (the limiting). Such consideration to him signifies all that is best in philosophy and it is that which he calls the true method of philosophy, and the knowledge of the truth; in it he places the distinction between philosophers and Sophists. (Hegel Werke 19, 67/Hegel 1892ff., vol. II., 54)

The true philosopher or dialectician, differently from the Sophist, focuses on pure concepts or pure thoughts, whereby "pure" refers, as we have seen, to

---

15 On Hegel's view on the infinite see here Chapter 4.3.
16 On Hegel's interpretation of Plato's dialectic see Düsing 1990, 169–191, Rosen 1990, 153–168, Baum 1990, 192–208, Asmuth 2006, 125ff.

the fact that the philosophical consideration is moved by the sole aim of analysing the concept, finding its true meaning, and by nothing else.

Hegel observes that Plato himself "did not show with sufficient clarity how [sophistic] is to be distinguished from the purely dialectical knowledge" (Hegel Werke 19, 71/Hegel 1892ff., vol. II., 63). In other words, Plato also practiced dialectics in the sophistic sense of a technique aimed at giving arguments and counterarguments. At the same time, however, he also repeatedly expressed his dissent from sophists. Thus the crucial logical question is to draw a distinction between the sophistic "making fleeting all that is secure" and the dialectical "showing contradictions within pure thoughts".

The difference is related to the notion of contradiction. In a much-discussed passage of the *Sophist* (259) Plato distinguishes between dialectical and sophistic treatment of contradictions. Hegel comments:

> Plato objected to this unity of opposites, because it must thereby be said that *something is one in quite another respect in which it is many* [my emphasis]. We thus do not bring these thoughts together here, for the conception and the words merely go backwards and forwards from the one to the other; if this passing to and from is *performed with consciousness* [my emphasis], it is the empty dialectic which does not really unite the opposites. (Hegel Werke 19, 63/Hegel 1892ff., vol. II., 50)

Hegel distinguishes here between sophistic and dialectical "unities of opposites" (contradictions) as conjunctions of opposite determinations (such as being and non-being, one and many, sweet and non-sweet etc.). The sophistic unities of opposites are not true contradictions because "something is one [or sweet, or large etc.] in another respect in which it is many [or not sweet, not large etc.]". The clue is in the expression "in another respect", which recalls Aristotle's formula *at the same time and in the same respect* (Aristotle De Interpretatione 6, 17, 34–37 – translation from Barnes 1984). Sophistic contradictions for Hegel are not true contradictions. They are simply *separate* assertions of contradictories: one person says that p and she is right *in her own terms*, another person says that not-p, and she, *in her own terms*, is right again.

On this point, Hegel reproduces what Plato himself writes in *Sophist* 259:

> What is really difficult and true is this: to show that what is the other is the same, and what is the same is another, and that in the same regard and from the same point of view [...] To show that somehow the same is another, and the other also the same, that the great is also small and the like also unlike, and to delight in thus always proving opposites, is no true insight (*elenchos*)[17] but simply proves that he who uses such arguments is a neophyte [in

---

17 Hegel places the Greek word *elenchos* (which literally means "refutation") beside the German

thought], who has just begun to investigate truth. (Hegel Werke 19, 72/Hegel 1892 ff., vol. II., 64)[18]

What Hegel wants to stress about Plato's passage in the *Sophist* is the positive upshot of dialectical refutations:[19]

> Thus Plato expressly speaks against the dialectic of showing how anything may be refuted from some point of view or another. We see that Plato [...] expresses nothing else than what is called indifference in difference, the difference of absolute opposites like the one and the many, being and non-being, and their unity. To this speculative knowledge he opposes the ordinary both positive and negative way of thinking. (Hegel Werke 19, 72 f./Hegel 1892 ff., vol. II., 64)

True insight is for Hegel the mark of *speculative* knowledge, that is, of the thought which brings absolute opposites (e. g. contradictories) together. This

---

translation, evidently because he wants to stress the technical meaning of the term Plato is using, and the fact that the German translation "Einsicht" does not refer to a generically intended "insight", but to the one resulting from an *elenchos* ("refutation"), or to a special form of refutation. That Hegel is well aware about the technical meaning of the word is evident in several passages of his works. In the chapter on Aristotle in the *Lectures on the History of Philosophy* he writes: "The treatise on *Sophistic Elenchi* [...] or "on sophistic refutations" [deals with] the ways in which the contradiction is produced in common thinking [*Vorstellen*] [...]. The sophistic *elenchi* betray the unconscious representation into such contradictions and make it conscious of them". Hegel Werke 19, 236–237/Hegel 1892 ff., vol. II., 218. The aim of the treatise is, as a matter of fact, to focus attention on the fallacies of common thinking (*Vorstellen*), rendering it aware about itself and its own mistakes. In the *Science of Logic* Hegel refers to the Megarian paradoxes, writing "they are familiar under the names of 'the bald' and 'the heap'. These *elenchi* are, according to Aristotle's explanation, ways in which one is compelled to say the opposite of what one had previously asserted" (Hegel Werke 5, 397/Hegel 1969, 335).

18 Hegel is here quoting Plato's *Sophist* 259, using the *Editio Bipontina*, based on Ficinus' Latin translation – see on this Düsing 1990, 182. The translation of this passage and Hegel's interpretation has been discussed. Eduard von Hartmann claimed in 1868, 8 that Hegel illegitimately identifies his own dialectics (his view about the truth of contradictions) with Plato's dialectics, on the basis of a "single obscure and disputed passage of the Sophist (*Sophist*, 259), which, in whatever way you might construe it grammatically, will at any rate exclude the Hegelian interpretation". For von Hartmann Plato thus did not want to claim that contradictions are true. Against Hartmann Michelet 1871, 322 writes: "It is incredible that [the passage] should still appear to him [Hartmann] obscure and doubtful, which has never been nor can be to any one possessing even but a fair knowledge of Greek". More recently some authors, among them Düsing 2012, 84 and Verra 2007 claim, not differently from von Hartmann, that Hegel's translation rests on a misinterpretation, and that Plato, in the afore mentioned passage, did not want to defend the "Hegelian" contradiction.

19 See on the mysterious positive addition here Chapter 12.

kind of thought manifests a peculiar attitude toward contradictions, different from what Hegel calls here "the positive and negative way of thinking". The positive attitude implies stating a thesis and *also* its negation, but separately, without even seeing (or pretending not to see) that they are contradictory. The negative one recognises the contradiction, but introduces different perspectives and parameters, without truly unifying the contradictories.

Accordingly, Hegel recalls that true dialectic is to be intended as

> [S]howing the necessary movement of pure concepts, without thereby resolving these into nothing; for the result, simply expressed, is that *they are this movement* [my emphasis], and the universal is just the unity of these opposite concepts. We certainly do not find in Plato a full consciousness that this is the nature of dialectic, but we find dialectic itself present. (Hegel Werke 19, 62/Hegel 1892ff., vol. II., 49)

The dialogue *Parmenides* is for Hegel the clearest example of genuine dialectic. Hegel quotes Plato:

> For example, in the case of the hypothesis "the many is" you have to consider what will be the consequences of the relation of the many to itself and to the one

And comments:

> It will become the opposite of itself; the many turns into the one insofar as it is considered in the determination it has. This is the marvellous fact that meets us in thought when we take determinations such as these by themselves, is that each one is turned into the reversal of itself. (Hegel Werke 19, 80/Hegel 1892ff., vol. II., 57f.)

Hegel stresses that dialectic implies taking determinations *by themselves*. In the Platonic dialogues dialectic is the method of the philosophical consideration, which implies asking: "what is X?" or "is X true?", and applying the concept which is to be analysed (being, the one etc.) to itself, asking "is the one one?", "what *is* being?". Hence we can say that "taking determinations in themselves" involves a semantic ascent from "simply talking" or "simply thinking" to "talking about talking" or "thinking about thinking", making the concepts or words we use the objects of our inquiry. As Hegel writes, the "dialectical consideration for Plato is consideration of what is to be taken as determination" (Hegel Werke 19, 81/Hegel 1892ff., vol. II., 57). And "dialectic is nothing else than the activity of thought thinking about itself" (Hegel Werke 19, 82/Hegel 1892ff., vol. II., 60). This is also the meaning of "pure thought" we have seen considering Zeno's method of refutation.

Thus if we reflect about the concepts involved in what we say, asking what they are, or if they are what they claim to be etc., then a "marvellous fact" meets us: the conceptual determinations imply (or turn into) their opposites.

> In the sentence "the one is" is implied "the one is not one, it is many" and conversely "the many is" simultaneously implies "the many is not many, it is one". They manifest themselves as dialectical, they are, essentially, the identity with their negations; and this is their truth. (Hegel Werke 19, 82/Hegel 1892ff., vol. II., 59f)

As an example Hegel mentions the concept of becoming:

> [I]n it being and non-being are contained; the truth of both is becoming; it is unity of both as indivisible and yet distinct, since being is not becoming and non-being is also not becoming. (Hegel Werke 19, 82/Hegel 1892ff., vol. II., 60)

Thus with becoming we have a complex concept that is internally contradictory, and different from its internally contradictory elements. It is different from them because it is the unity of the two elements and its description is not exhausted by only one of these determinations. The meaning of "becoming" is thus exhausted neither by "being is" nor by "being is not", neither by "being is nothing" nor by "being is not nothing". Rather it is fully expressed only by the "unity" of the two. As I will explain, this idea about the inseparable unity of contradictories is a crucial principle of Hegel's logic.

10.3 The Megarians

The Megarians, Hegel says, practiced dialectic "in a kind of anger, so that others said that they should not be called a school (*scolé*) but a gall (*colé*)", though Euclides, the founder of the school, "in spite of his stubborn manner of disputing [is said] to have been, even in his disputation, a most peaceful man".[20] Once in a discussion his opponent was so irritated that he exclaimed: "I will die if I do not revenge myself upon you!" and Euclides replied: "and I will die if I do not soften your rage so much by the mildness of the grounds (*lenitate verborum*) that you will love me as before". In the Megarian practice we find the discussive (pragmatic) meaning of dialectic in its clearest expression. The other aspects of dialectic are also at work.

---

20 On Hegel's interpretation of Megarian paradoxes see d'Agostini 2008, Chapter 12 and d'Agostini 2011b, 121–140.

The Megarians held that only the universal (the Eleatic "one", the Socratic and Platonic ideas) is true. Accordingly, their dialectic consists in showing that "all that is determined and limited is not true" and "in bringing all that is particular into confusion and annulling this particular". In this respect the Megarian dialectic is very close to the Eleatic one, and brings it "to very great perfection". Its aim is to reveal the primacy of the universal (the conceptual) in the human search for truth, and consequently in human reasoning (Hegel Werke 18, 523/Hegel 1892ff., vol. I., 454).

Yet Hegel also stresses that the Megarians somehow connected the Eleatic (and Socratic) results about universals to the sophistic practice:

> With a dialectic thus constituted, we find them taking the place of the Eleatic School and of the Sophists. (Hegel Werke 18, 523/Hegel 1892ff., vol. I., 454)

In Hegel's reconstruction, Megarians inherit from the sophists the idea of conceptual movement (that the semantics of concepts involves contradictions), while from the Eleatics they inherit the idea of the unity, uniformity and completeness of being.[21] Hence the Megarian dialectic, in distinction from sophistic practice, does not aim at the destruction of concepts, but "to simple universality as fixed and as enduring".[22]

Thus Megarian dialectic brings, similarly to sophistic dialectic, all that is particular into confusion, showing its contradictions, though it does so with the aim of establishing the truth of the universal.[23] This is precisely, from Hegel's

---

[21] In this respect, Megarian dialectic is close to Eleatic dialectic, but identifies the universal with the (Socratic) good, and not primarily with the one. The Megarian Euclides was, in fact, famous for his view according to which "the good is one, and it alone is, though passing under many names; sometimes it is called understanding, sometimes God; at another time thought (*nous*), and so on. But what is opposed to the good does not exist" (Hegel Werke 18, 524/Hegel 1892ff., vol. I., 454–455).

[22] d'Agostini 2008a, 203–204 reconstructs Hegel's account as follows: "First the Eleatics discovered a specific dialectics concerning being and Zeno reveals the contradiction involved in the nexus between being and other concepts, such as movement and plurality. Second, with the sophists the result is *generalised*. The sophists realised that each concept can be treated in the same way: contradictions can be drawn from everything, from every conceptual content. Third the Megarians insert the Socratic element into this picture. They adopt the generalisation of contradiction discovered by the sophists, but with a new awareness concerning universals (concepts), and keeping to the idea that dialectics – the art of contradiction – is not mere rhetorical joke but a method for grasping philosophical truth".

[23] As d'Agostini underlines: "The Megarians have in common with the Eleatics, and then with Zeno, the use of contradiction in defence of truth. The contradiction is used in both cases to reduce to absurdity the unacceptable theses of the *doxa*, and to defend philosophical *aletheia*. But

point of view, the ultimate aim of the paradoxes developed by Eubulides (the Liar paradox, the Concealed, the Sorites among others).

Megarian paradoxes are especially important for understanding the nature of Hegel's dialectic in the three aspects of discussive practice, semantics of higher order concepts and logic of contradiction, as well as for understanding how these three aspects converge in Hegel's making of dialectical inferences the eminent form of valid inference.

The paradoxes were expressed in the form of questions, as was usual in ancient logic: "did you stop beating your father?" or: "if a man acknowledges that he lies, does he lie or speak the truth?" (Hegel Werke 18, 531/Hegel 1892ff., vol. I., 459). Their paradoxality was not based on the unacceptable nature of the conclusion of a sound argument, as in the canonical contemporary definition of paradox,[24] but on the fact that these sorts of questions require a double answer: yes and no. Hegel comments that a simple answer is demanded, since "the simple whereby the other is excluded, is held to be the true". But a simple answer, such as "yes" in the case of "did you stop to beat your father?" or "the man tells the truth" cannot be given. As a matter of fact, "yes" means: "I once beat him" and "no" means: "I still beat him". Similarly, in the case of the Liar paradox:

> [I]f it is said that he tells the truth, this contradicts the content of his utterance, for he confesses that he lies. But if it is asserted that he lies, it may be objected that his confession is the truth. He thus both lies and does not lie; but a simple answer cannot be given to the question raised. For here we have a union of two opposites, lying and truth, and their immediate contradiction. (Hegel Werke 18, 529/Hegel 1892ff., vol. I., 459f.)

Eubulides requires that his opponents answer in a simple way, saying either "yes" or "no", either "the Liar tells the truth" or "he lies". In so doing, he follows the principle of ordinary logic according to which it is not possible that something is and is not, or that a sentence is true and false at the same time (the Law of Non-Contradiction: LNC) and the principle according to which something is either true or false, and there is no third possibility (The Law of Excluded Middle: LEM, or, as Hegel calls it, the *principium exclusi tertii*). Eubulides' aim was to show that the content of these paradoxes is contradictory, that they break the "law of the simplicity of truth", and are therefore false. Hegel stresses instead that paradoxes are "truly contradictory", that is, they show that the truth is

---

Megarians do not limit themselves to removing 'plurality' and 'movement' [...] rather, they globally reduce to absurdity common language [...] Every particular conception is disproved, in favour of the pure identity of the universals" (d'Agostini 2008a, 204).

24 See Sainsbury 2009, Introduction.

not (always) simple, and that LNC and LEM are not sufficient to give an account of truth. He writes:

> Menedemus hence replied that he neither ceased to beat him, nor had beaten him; and with this his opponents were not satisfied. Through this answer, which is two-sided, the one, as well as the other, being overcome [*aufgehoben*], the question is in fact answered; and this is also so in the former question as to whether the man spoke truly who said he lied: he speaks the truth and lies at the same time, and the truth is this contradiction [...] These sophisms thus not only are the appearance of a contradiction, but real contradiction is here at stake. In the example two things are set before us, a choice, but the determination is itself a contradiction. (Hegel Werke 18, 531/Hegel 1892ff., vol. I., 461)

The Liar paradox is especially relevant for explaining the nature of dialectical inferences and the reasons of their validity. The double truth conveyed by paradoxical self-reference clearly points the way to the idea of the internally contradictory nature of higher order concepts. I will review all of this in detail in the last chapter of this part.

10.4 Aristotle

Hegel complains in the preface to the *Phenomenology of Spirit* about the "separation of dialectic from philosophical proof".[25] Dialectic for Hegel is the method of philosophical demonstrations. In this he declaredly follows the Ancient Greek rather than the modern tradition. Some authors have taken this claim to mean that Hegel adopts Socrates' and Plato's, rather than Aristotle's view, but this interpretation is reductive and misleading. First, many authors, first of all Berti 2015, underline that there is open continuity between the Socratic-Platonic and the Aristotelian theories of dialectic.[26] Second, Hegel's observations in the

---

[25] On Hegel's interpretation of Aristotle's logic see Mignucci 1995, 29–50. On Hegel and Aristotle see Berti 1990, 65–83, Düsing 1990, Aubenque 1990, 208–226, Ferrarin 2001, Dangel 2013. Pöggeler 1970, 307 claims that Hegel und the Hegelians (first of all Michelet) "undo what Aristotle had achieved in his *Organon*", namely "the clear distinction between *apodeixis* and dialectic. The meaning of dialectic as topics, the connection of arguments to argumentative recommendations, and to specific positions in the open, historical dialogue gets lost". Pöggeler's view is almost unanimously shared by the interpreters, among others by Gadamer 1976, Aubenque 1990, 208–226, Baum 1986, 6–29 and Dangel 2013, 17ff.

[26] In the *Topics* Aristotle stresses the demonstrative, syllogistic dimension of dialectic, so that many interpreters share the view that dialectic can plausibly be seen, in Aristotle's terms, as the method of the (philosophical) science. See Berti 2015, 138–146 as well as Rapp/Wagner 2004, 36. In the presentation of Aristotle's position I follow Berti 2015, 9ff. and 107ff. According to

Lectures on the History of Philosophy show that he is well aware of the scientific dimension of Aristotle's dialectic. Third, and more generally, we have seen that, as far as logic is concerned, there is clear accordance between Hegel's and Aristotle's conception.

While Hegel, as we have seen, extensively discusses the general meaning of logic emerging from the *Organon*, he does not devote longer discussions to the meaning of dialectic in Aristotle. With reference to Aristotle's treatment of Zeno's dialectic of motion and change, he writes that:

> Aristotle solves it through the universal. He says that [motion and change] are this contradiction, they are what contains the opposition in itself, the universal; their unity, in which their moments dissolve, is not a nothing. [Aristotle does not state that]: motion and change are not; but [rather that they] are a negative and a universal. (Hegel Werke 19, 192/Hegel 1892 ff., vol. II., 174)

Aristotle's position is so seen as a prefiguration of Hegel's own view on Zeno's dialectic of motion. Zeno's demonstrations, according to Hegel, do not refute the existence of empirical motion, but make the internally contradictory nature of the *concept* of motion explicit.

In the context of his discussion of Aristotle's logic, Hegel briefly refers to the refutations in *Sophistic Elenchi* as methods in order to produce contradictions, and thus make the person who contradicts herself *aware* of her own thought. *Elenchi* are, in this light, instruments to produce the passage from logical unawareness to logical awareness, from our unconscious using of forms of thought to our making them the object of our inquiry.

But it is in the *Topics* where the Aristotelian theory of dialectical inferences is fully deployed. Hegel explains that "topics" are

> [T]he points of view from which anything can be considered [...] Aristotle gives a large number of general points of view which can be taken of an object, a proposition or a problem; each problem can be directly reduced to these different points of view, that must everywhere appear. Thus these "places" are, so to speak, a system of many aspects under which an object can be regarded in investigating it [...] the knowledge of points of view at once places in our hands the possibility of arriving at the various aspects of a subject, and embracing its whole extent in accordance with these points of view. (Hegel Werke 19, 235 f./Hegel 1892 ff., vol. II., 217 f.)

Berti, the reading that underlines the merely negative meaning of dialectics in Aristotle and sees dialectics as a discipline aiming at testing the discourses of others, rather than achieving new knowledge, is insufficient. Aristotle, for Berti, understood dialectic as the logic of philosophy. In philosophy dialectics is not opposed to *apodeixis*, but corresponds to it. Differently from Berti, I stress the continuity between Aristotle's and Hegel's conception of dialectic.

This means that the books *Topics* coincide with:

> [D]ialectic – external determinations of reflection. Aristotle says it is an instrument for finding propositions and conclusions out of probabilities [...] he says that we must use syllogisms with the dialecticians, but inductions with the multitude. In the same way Aristotle separates the dialectic and demonstrative syllogisms from the rhetorical and every kind of persuasion; he counts induction as belonging to what is rhetorical. (Hegel Werke 19, 236/ Hegel 1892ff., vol. II., 218)

Dialectical inferences for Aristotle are thus deductive (syllogisms), and their premises are probable. The premises of dialectical inferences are, more specifically, *endoxa* – theses defended by the majority of people or by the experts expressing controversial views about, among other subjects, the same principles and concepts of philosophy (such as the one, being, truth etc.). For Hegel, the dialectical illustration in the *Topics* deals with "external determinations of reflection".

Interpreters (Berti 2015, 138 ff. and Rapp/Wagner 2004, 7 ff.) agree on the fact that the *Topics* contain Aristotle's systematic presentation of the dialectical method Plato developed in the *Parmenides*. For both Plato and Aristotle dialectic has, among other uses, a fundamentally philosophical use. In other words, as Aristotle says, it is useful in order "to develop aporias in both directions", a development which, according to him, "produces a genuine knowledge of truth" and helps "to detect both the true and the false" (Aristotle, *Topics* I 2, 101 a 34 b4 – translation from Barnes). Similarly, Hegel answers Goethe's question "what is dialectic?", as we have seen, by stating that "it is nothing else than the systematic development of the spirit of contradiction, which is essential in order to distinguish truth from falsity" (Eckermann 1987, 622–623). In the third book of the *Metaphysics* Aristotle presents the development of the aporias as the same method of philosophy:

> It is necessary that in relation to philosophy we find first the things about which the aporias are to be posited: these are the things on which there are disagreeing views [...] for those who want to solve the apories (*euporesai*) it is first necessary to develop the aporias well (*to diaporesai kalos*), since the further proceeding (*e usteron euporia*) depends on the solution of the aporias previously posited (*ton proteron aporoumenon*). (Aristotle, *Metaphysics* III 1, 995 a 24–29 – translation from Barnes 1984)

As Berti suggests, Aristotle presents here the different passages immanent to every inquiry into "first principles" or basic philosophical concepts: advancing the opposition (the one is one – the one is not one), then developing it, that is deriving the consequences from each part of the contradiction, and finally solving it. Berti writes:

> It is evident that here [Aristotle] applies the law of non-contradiction as criterion in order to detect the false and the principle of excluded middle as criterion in order to detect the true. But in order for such a discovery to take place it is necessary that the field of views under scrutiny exhausts all possible views, that it produces an alternative between contradictory and not merely contrary propositions. (Berti 2015, 139)

The continuity between the Aristotelian and Platonic procedure in the *Parmenides* (and the Hegelian conception of dialectic) is thus evident. While dialectic in Aristotle is a technique in order to find truth, in Hegel it is both the method and the nature of the inquiry into truth. In this sense it is the *form* of (the search for) truth, and the form of (the search for) truth is the specific *content* of Hegel's philosophical logic.

10.5 The Sceptics

"Scepticism" Hegel declares in the *Lectures on Logic and Metaphysics* "is the logical moment of dialectics" (Hegel 1992, 14).[27] The sceptics developed a peculiar awareness about the forms of reasoning, different from the traditional logical one. More specifically, they demonstrated "contradictions through the tropes".

> These tropes prove that the Sceptics had a deep awareness about the process of argumentation – a deeper one than is found in ordinary logic, the logic of the Stoics and the canon of Epicurus. These tropes are necessary contradictions into which intellectual thought [*der Verstand*] falls. (Hegel Werke 19, 394/Hegel 1892ff., vol. II., 365)

Hegel writes that the sceptical procedure of showing contradictions in every determinate thought requires abstraction, as well as awareness about the forms of thinking and reasoning:

> [T]o acknowledge the forms of opposition everywhere, in every concrete material, in every thought, requires a clear force of abstraction [...] Two formal moments are typical of the sceptical way of thinking: a) the ability to force ourself to become aware about our own operations, making them the object of consideration b) we state a sentence, are normally concerned with its content [...] commonly we do not know anything about it but for this content, we do not know anything about its form [...] the sceptics do not fight for the content, but rather grasp the essence of what is said, the whole principle of assertion. (Hegel Werke 19, 395 f./Hegel 1892ff., vol. II., 365 f.)[28]

---

[27] On Hegel's interpretation of ancient scepticism see among others Buchner 1990, 227–243, Varnier 1987, 282–312, Vieweg 1999 and 2007, Heidemann 2007.
[28] I have partially changed the translation to bring it closer to the letter of the German text.

Thus scepticism is, for Hegel, a formal knowledge, a knowledge about "the whole principle of assertion", which means that the sceptics developed a formal way of thinking, focused on the most general aspects of our thoughts rather than on their contents. Considering thoughts, the Sceptics were principally interested in relations of thoughts to each other, or in the affirmative or negative expression of sentences, rather than their content. This formal moment parallels "taking conceptual determinations *per se*" that is typical of Platonic dialectic, insofar as it involves some sort of semantic ascent: the act of becoming aware of our own operations (which Hegel would call reflexivity).

The sceptical tropes are, very simply, forms (or kinds) of arguments that show the irrelevance or inaccessibility of truth. Thus we have here forms employed against truth. The ultimate aim of the tropes is to confirm the sceptical principle which, in its basic formulation, states: "for every valid argument there is an opposite one that is equally valid" "for every true sentence there is an opposite one that is equally true". This is the reason why, as Hegel writes, the Sceptics "acknowledge the forms of opposition everywhere, in every concrete material, in every thought" and this "requires a clear force of abstraction". The two formal movements specified above: a) reflexivity and b) independence of content, interestingly, coincide with the basic moves that open the field of *Vernunftlogik*.

In the *Essay on Scepticism and Philosophy* Hegel deepens the logical meaning of scepticism, and its relation to *Verstandeslogik* (the formal logic of Hegel's times) and *Vernunftlogik* (dialectical logic). In particular, Hegel's analysis of the later five tropes (opposition, *diallele, regressus ad infinitum*, presupposition, relativity) is important from a logical point of view.[29] As a matter of fact, while the earlier tropes are directed against sense-certainty, the later ones show the limits of *intellectual* (logical and metaphysical) knowledge (*Verstandeslogik* and *Verstandesmetaphysik*). The five tropes of Agrippa are "necessary contradictions into which intellectual knowledge falls". In this respect, they are a prefiguration of Hegel's own rational logic and metaphysics (*Vernunftlogik*). However, sceptical tropes have with respect to *Verstandeslogik* a merey negative and confutative function, while the Hegelian perspective of *Vernunftlogik*, though entailing the sceptical moment, has a specifically positive and foundational relevance.

For Hegel, however, the sceptical account does not capture the logical significance of dialectic as form of valid reasoning, for two main reasons. First, scep-

---

[29] Verra was among the firsts to highlight the fundamental importance of the ancient sceptical tropes for Hegel's dialectical method. See now Verra 2007, 55–64. See also Varnier 1990, Buchner 1990, 227–243 and Vieweg 1999.

ticism remains at the negative moment. Second, the sceptics were not aware of what they, ultimately, were doing and thus lacked sceptical self-awareness. Let us examine these two aspects in more detail.

As to the first point, Hegel writes:

> [Scepticism] is the dialectic of all that is determinate [it shows the limit] of every [simple] truth, [insofar as it shows] that it contains a contradiction [...] the logical concept is itself this dialectic of scepticism, for this negativity which is characteristic of scepticism likewise belongs to the true knowledge of the idea. The only difference is that the sceptics remain at the result as negative, saying: 'This and this has an internal contradiction, it thus disintegrates itself, and consequently does not exist'. (Hegel Werke 19, 360/Hegel 1892ff., vol. II., 330)

The passage recalls the distinction between Zeno's (negative) and Herlaclitus' (positive) dialectic. The sceptical dialectic is negative in that it states: "this and this has an internal contradiction, and thus disintegrates itself". From the sceptical point of view, the tropes simply show the inaccessibility and irrelevance of truth.

As to the second point, while intellectual thought is easily attackable by scepticism (by one or a combination of the five tropes), Hegel says that the five tropes fail when applied to rational and speculative contents.[30] In the *Essay on Scepticism and Philosophy* Hegel explains this failure as follows:

> If applied against dogmatism these tropes are rational, because for a [determination presented by the dogmatic philosopher] they let emerge the opposite one, from which [the dogmatic philosopher] abstracts away – in so doing they produce the antinomy. But if applied against reason, they maintain the mere difference which affects them; the rational element they have is already contained in rational thought. (Hegel Werke 2, 246)

In other words, if used to refute intellectual thought, the sceptical tropes are rational, if used to attack reason, they are intellectual (dogmatic) and thus self-refuting. As Hegel more figuratively puts it: "um [das Vernünftige] kratzen zu können, [diese Tropen] geben ihm die Kratze der Beschränktheit" which means "in wanting to 'scratch' reason, the tropes give to it the scratch of limitation" (Hegel Werke 2, 247).

To see all this, it is useful to consider the example presented by Hegel himself in the *Lectures on the History of Philosophy* and in the *Essay on Scepticism and Philosophy*. Hegel refers hereby to Sextus' refutation (in *Adversus mathema-*

---

[30] "These forms" Hegel writes in the *Lectures on the History of Philosophy* "do not apply to what is speculative". Hegel Werke 19, 398/Hegel 1892ff., vol. II., 364.

ticos) of Aristotle's view about the idea of *noesis noeseos* (thought thinking about itself/reason comprehending itself). The concept of *noesis noeseos* is the paradigm of rational thought, one could say that it is the very definition of rational thought. Sextus argues that:

> [T]he reason that comprehends is either the whole or it is only a part. If reason as the comprehending is the whole, nothing else remains to be comprehended. If the comprehending reason is, however, only a part which comprehends itself, this part again, as that which comprehends, either is the whole (and in that case again nothing at all remains to be comprehended), or else, supposing what comprehends to be a part in the sense that what is comprehended is the other part, that which comprehends does not comprehend itself. (Hegel Werke 19, 399/Hegel 1892ff., vol. II., 369)

Hegel objects that Sextus

> [B]rings into the relationship of thought thinking about thought the very superficial category of the relationship of the whole and the parts [but] the relationship of whole and part is not a relationship of reason to itself. (Hegel Werke 19, 400/Hegel 1892ff., vol. II., 370)

The passage may sound obscure, but it has a simple meaning that is logically relevant to understand the notion of dialectical (rational) validity Hegel had in mind.

Sextus presupposes that "reason is *either* the whole *or* the part". Here Hegel objects that

> [T]o the knowledge of what is speculative [i.e.: of reason] it belongs that there is, beyond the either-or, a third: it is the "both...and" and "neither ... nor". (Hegel Werke 19, 399/Hegel 1892ff., vol. II., 369)

In other words, the Law of Excluded Middle does not hold for the properties "being a whole" and "being a part" applied to "rational thought". As a matter of fact, rational thought is self-referential thought. Hence it is by definition both a "part" (the act of thinking, different from what is thought in such an act) and at the same time the "whole" (since what is thought is the same act of thinking). In other words, reason has a special way of being part and being whole.

Finally, it should be noted that "formal" for Hegel means self-referential, as we have specified above (especially in Part II). In this regard, the sceptics' formal approach betrays itself. Hegel recognises that the sceptical position is untenable if considered from a formal point of view, namely if one forces the sceptic to reflect about her/his own position:

It is this formal appearance of a sentence with which the Sceptics are usually bulled around, insofar as one gives them back what they give to him/her, namely that even if they doubt about everything, *that* they doubt is certain [...] In this extreme case of highest consequence [...] scepticism had to become inconsequent, since (one) extreme cannot persist without its opposite. (Hegel Werke 2, 248–249)

The "extreme consequence" here consists in applying scepticism to scepticism. The product of this self-application is said to be "inconsequence", as through this self-application the sceptic is forced to assume the opposite of what s/he states. In this respect, this passage also shows the specifically Hegelian standpoint about dialectical refutations. The application of the sceptical (formal, second order) perspective to scepticism itself is not properly a *refutation* of scepticism, but rather its *completion*, and transmutation into dialectical thought. In other words, "giving the sceptics back what they give to you" implies showing that the sceptical thesis "everything can be doubted" entails its negation, namely: "not everything can be doubted".

Notably from these sorts of implications with the form "if A then ¬A" Hegel does not infer "¬A" but rather "A ↔ ¬A". In the last chapter of this part, I will go back to this Hegelian inference by considering its meaning for the history of logic and for philosophical logic.

10.6 Kant

At the end of his consideration of Zeno's dialectic Hegel says:

> This is the dialectic of Zeno; he considered the determinations which our ideas of space and time contain, and showed in them their contradiction; Kant's antinomies do no more than Zeno did here. (Hegel Werke 18, 317/Hegel 1892ff., vol. I., 277)[31]

For Hegel, Kant's dialectic corresponds to Zeno's dialectic insofar as it is a consideration "of the determinations our ideas contain, which shows their contradictions". The antinomies are theses about the world, which is, as Hegel emphasises, a universal concept. Kant shows that, as soon as we try to determine its nature, we find contradictions. Kant's and Zeno's dialectic implies purity (that is: "putting oneself right into the thing", analysing the concept of space, time,

---

[31] On dialectic in Kant and Hegel see among others Wolff 1981, Düsing 1983, Verra (ed.) 1981, Baptist 1986, Priest 2002, 102ff., Brinkmann 1994, 57–68, Engelhard 2007, 150–170, Sedgwick 2012.

the world, abstracting from everything else), and the idea that contradictions naturally follow from this pure analysis.

The difference between the Kantian and the Eleatic approach is described by Hegel as follows:

> Though the content is, also in Zeno, null, in Kant [this happens] because of our intervention. In Kant it is our thought that ruins the world [...] it is our application, our input alone that ruins it, what we do on it is good for nothing [...] in Zeno the world is in itself appearance and untrue. (Hegel Werke 18, 318/Hegel 1892ff., vol. I., 277)

Both Zeno and Kant show contradictions within concepts, i.e. the concept of the world, matter, motion etc. However, while for Kant the thinking subject is the cause of the emergence of contradictions, and what is thought (the world, matter etc.) remains in itself intact, for Zeno the same concept of a sensible world breaks up and turns out to be false. In this sense Hegel stresses that "Zeno's dialectic has a greater objectivity than this modern dialectic".

In the passage of the *Lectures on the History of Modern Philosophy* on Kant's antinomies Hegel stresses that Kant, in the *Critique of Pure Reason*,

> [P]oints out four contradictions. But this is too little, antinomies are everywhere. It is easy to show a contradiction in every concept, because the concept is concrete, not a simple determination. Thus it contains different determinations, and these are at the same time opposites. (Hegel Werke 20, 356/Hegel 1892ff., vol. III., 448)

The contradictions can be found in every concept "because the concept is concrete" and this means that concepts are heterogeneous structures, involving different aspects, and for Hegel, as we will see better in what follows, differences are (or turn into) contradictions. The concepts analysed by Kant in the four antinomies are for Hegel universal or higher order concepts:

> The antinomy is the contradiction [emerging from] reason's idea of the unconditioned, an idea applied to the world in order to represent it as a complete summing-up of conditions. That is to say, there are phenomena, and reason demands the absolute completeness of the conditions of their possibility, [these conditions] constitute a series, and reason requires a thoroughly complete synthesis of these conditions. If now this completeness is expressed as existing, only an antinomy is presented, and reason is presented only as dialectical. In this object there is in every respect a perfect contradiction. (Hegel Werke 20, 356/ Hegel 1892ff., vol. III., 448)

The antinomy for Kant is the contradiction emerging when we express the infinite series of the conditions as an object. The series is infinite and the object is as such finite, one single object different from others. Thus the same contradiction we saw considering Hegel's interpretation of the fraction 2/7 as an expression of the good infinite, i.e. a finite sum expressing an infinite content (the infinite series of the numbers) emerges.[32]

Hegel praises the Kantian analysis for the reason that

> [O]ne of these opposites is just as necessary as the other [...] The necessity of these contradictions is the interesting fact which Kant has brought to consciousness; in ordinary metaphysics, we think that one must hold good, and the other be disproved. But the necessity that such contradictions happen is precisely what is interesting. (Hegel Werke 20, 358/ Hegel 1892ff., vol. III., 450)

The Kantian dialectic for Hegel is genuinely "interesting". In fact, in it the contradictions are *necessary*.

---

[32] See 4.3.

11 Hegel's own account of dialectical inferences

In the last paragraphs of the "Preliminary Considerations" in the *Logic* of the *Encyclopaedia* Hegel presents the famous triadic development of dialectical thought as "the general structure of the logical [das Logische]" and "rule of every development of the idea" (Hegel Werke 8, 168/Hegel 1991, 125). In my view, this passage is important for illustrating Hegel's general view on validity. By "idea" is meant, as we have seen, a kind of thought involving a coincidence between reality and rationality, i.e. the kind of reasoning that is able to grasp things as they are, or, in short, true thought – in the philosophically relevant Hegelian sense of *Wahrheit*, distinguished from *Richtigkeit*. Hence "the rule of every development of the idea" is simply the form of truth-conveying arguments. As we have seen, and will elaborate further later on, for Hegel validity as the necessary passage from premises to conclusions in arguments is to be rooted in and oriented by truth. Since the overview of Hegel's history of dialectic already gives all the important elements to introduce into his view on logical consequence, I limit myself here to a schematic account. Hegel presents the view as follows:

> The logical [das Logische] has thus, from the formal point of view, three sides: α) the *abstract* or *intellectual* one; β) the dialectical or *negative-rational* one; γ) the *speculative* or *positive-rational* one. These three sides do not make three *parts* of logic, but are *moments of everything that is logical-real [Logisch-Reellen]*, that is, of every concept, or of every truth whatever. (Hegel Werke 8, 168/Hegel 1991, 125)

The three sides are "moments of every concept", of everything that is true, of everything that is both thought and real (*Logisch-Reell*). As we have seen, truth-preserving and truth-conveying thought for Hegel is conceptual thought. "Having the concept" or "thinking conceptually" means, for Hegel, thinking truly, and this corresponds to thinking thoroughly and sceptically, having a complete and sceptical thought about something. Thinking truly means engaging in the process of analysing concepts and determining their adequate meaning. It means dealing with the Socratic question "what is X?" (what is matter? What is the one? What is justice?).

In this light we can take the quote above to mean that the "three sides" express the form of every true thought, i.e. of every truth preserving and truth conveying inference. Let us see them in more detail:

> α) Thought, as understanding [*Verstand*] sticks to the fixed determination and its difference from other determinations; every such limited abstract it treats as having a subsistence and being of its own. (Hegel Werke 8, 169/Hegel 1991, 125)

The first moment or "side", α), consists in answering the Socratic question "what is matter?" proposing a single conceptual determination, distinct from other determinations, and considering it the full expression of the content at stake. We determine for example the concept of matter stating: "matter is continuous" (*m*), and we stick to this one determination, considering it adequate to express the whole concept.

> β) The dialectical moment is the moment in which these finite determinations supersede themselves, and pass into their opposites. (Hegel Werke 8, 172/Hegel 1991, 128)

In the second moment, the meaning of dialectics comes into play. Hegel himself explains this passage recalling the difference between dialectic and both scepticism and sophistic we have already seen in considering the *Lectures on the History of Philosophy*. He writes:

> The dialectical [*das Dialektische*] as seen in isolation by the understanding, corresponds to scepticism; it contains the mere negation as the result of dialectics.
> But in its true and proper character, dialectic is the peculiar and true nature of the intellectual determinations [...] Reflection is first going beyond the isolated determination, and putting it in relation [to other determinations], through which it is related to others, but kept in its isolated validity. But dialectic is rather this immanent going beyond, in which the one-sidedness and limitation of the intellectual determinations presents itself as what it is, namely their negation. (Hegel Werke 8, 172/Hegel 1991, 128)

The passage may seem cryptic, but it is easily explained from the point of view of what Hegel, in the *Lectures on Ancient Philosophy*, calls "genuine and *immanent* dialectic with a positive result", and distinguishes from both scepticism and sophistic.

The "Dialectical [*das Dialektische*] as seen by the understanding" means that from the abstract point of view of the understanding contradictions are taken as sign of falsity. Their emergence implies that the content at stake, as Hegel points out in the *Lectures on the History of Philosophy* with reference to Zeno, "breaks up". As such, the intellectual approach to contradictions is identical to scepticism. As we have seen, in the *Lectures on the History of Philosophy* as well as in the *Essay on Scepticism* Hegel defines ancient scepticism as the insight into the contradictory nature of every *logos* (statement, thesis). Even if their view was right, the consequences ancient sceptics drew from it are, according to Hegel, fundamentally misleading. In Hegel's view, ancient scepticism was based on the insight that for every valid argument of reason "there is an opposite one which is equally valid" (*panti logo logos isos antikeitai*), and this insight led to the dismissal of any theoretical inquiry. Hegel holds instead that the awareness concerning the "necessary contradictions" of thought implies the individuation of philosophy's specific method. So, in Hegel's view, in trying to destroy reason, sceptics discovered the method of reason.[33]

Dialectics should also be distinguished from a mere "technique which arbitrarily produces confusion in particular concepts and a mere appearance of contradictions" (Hegel Werke 8, 172/Hegel 1991, 128). In the *Lectures on the History of Philosophy* Hegel called this technique, as we have seen, "sophistic". It is worth noting that the expression "immanent" occurs here in exactly the same meaning it had in the *Lectures* as the qualification of genuine dialectic, which is motivated by the sole aim of analysing the concept, and is distinct from the sophistic technique, which is motivated by external interests.

We can illustrate the meaning of the "second moment or side" by recalling the example mentioned above. The abstract, intellectual point of view does put the first determination (m) in relation to other determinations (and also to the negation of m), but sticks thereby to the conviction that the original determination is the true one. In contrast, the dialectical point of view recognises m as limited and insufficient if one wants to give a complete account of the concept of matter, and sees the limitation of m "for what it is", namely the negation of m: in other words, if "matter is continuous" does not give a full account of what pertains to the concept of matter, then the negation of m: "it is not the case that matter is continuous" has the right to be stated. Here we see that concepts for Hegel are heterogeneous in the sense that they include different and opposite determinations. Single statements expressing one determination are thus one-sided and simply insufficient. The typical dialectical move, criticised in the whole reception of Hegel's dialectic, starting from Trendelenburg and leading through to Croce and Adorno, consists in showing how differences or simply contrary relations included in our thoughts on the concept, such as the relations among the sentences: "matter is continuous", "matter is extended", "matter is made of atoms", "matter is punctual", turn into contradiction, i.e. into the dual relation between "matter is continuous" and "matter is not continuous". I will further consider the question concerning the reduction of difference or contrariety to contradiction in the last part of the book, examining the meaning of negation in Hegel's logic.

On the last "moment" or "side" Hegel writes that:

> γ) The speculative or positive-rational moment grasps the unity of the determinations in their opposition – the affirmative, which is entailed in their disintegration and in their transition. (Hegel Werke 8, 176/Hegel 1991, 131)

---

[33] See Hegel Werke 19, 359 ff./Hegel 1892 ff., vol. II., 329 ff. and Hegel Werke 2, 230 ff.

The last moment is what Hegel called in the *Lectures* the "positive result" of "immanent dialectics". If we consider the example mentioned above, the understanding determines the concept of matter as involving continuity, and thus as not involving other different and opposite determinations, such as punctuality. In so doing the point of view of the understanding is abstract, because it isolates one determination, i.e. continuity, from all other possible predicates pertaining to the fully developed concept of matter. In contrast, the rational or speculative point of view implies that what is different and even contradictory, i.e. what discards the original assumption on the content at stake, is seen as essentially pertaining to the whole meaning of the concept, and thus as genuinely connected with the original assumption. In this connection, Hegel sees the restoration of the concrete nature of concept, which was destroyed through the abstracting operation of the understanding.

In sum: first, conceptual determination are fixed and isolated from different and opposite ones. Second, they are put in relation to each other, and this relation reveals itself as destructive: the integrity of the concept is dispersed. Finally, the concept's wholeness is restored. In these passages the concreteness of the concept, i.e. its complete meaning, is achieved.

All this clearly shows how Hegel's analysis of the history of dialectic from Zeno to Kant merges into his own account of dialectic. What needs to be examined now is the specifically logical meaning of Hegel's general notion of validity. In short, my thesis is that the "three sides" of *das Logische* are inferential moves unfolded in dialectically valid inferences, and are revelatory as far as the meaning of dialectical validity is concerned.

12 What is dialectic?

The philosophical historiography concerning dialectic is immense and complex.[34] As Hintikka writes "dialectic has the tendency to multiply itself beyond

---

[34] On the relation between dialectical validity and classical validity see Gadamer 1976, Düsing 1976, Fulda 1978, 33–69, Kulenkampf 1970, Günther 1978, Marconi (ed.) 1979a, Priest 1989, 388–415. For a complete and critical appraisal of these and other canonical positions see Marconi 1979b, 20 ff. Gadamer 1976 underlines the paradoxical nature of dialectical inferences. Dialectics is for Gadamer analysis of conceptual determinations, and the conceptual determinations are words deposited in language. The paradox consists in the fact that logic cannot but following language, trying to fix it univocally, but the plurivocity of language makes the fixation impossible. This means for Gadamer that Hegel's dialectical project of an exhaustive account of the syntactic-semantic links between categories fails. On Gadamer's view about the link between logic and dialectic see also Marconi 1979b, 21. Similarly, Kulenkampff 1970 describes dialectics as the attempt of formulating in the language the semantic properties of the language, i.e. of constructing a semantically closed system. In Tarskian terms, this means that the language contains its own truth-value, and this is the reason why contradictions emerge, for Kulenkampff. On Kulenkampff's analysis as well as its limits see Marconi 1979b, 24 f. Fulda 1978, 33–69 interprets dialectics as the process of making the initially implicit, vague (and perhaps wrong) assumptions about the conceptual terms explicit and more precise, and of correcting them. These assumptions are sentences of the form the "t1 is the t2", whereby t1 and t2 are called by Fulda "*Interpretamente*", by which Fulda means the theme of our discourse and interpretation. Consequently, the "is" means that "the theme of a discourse, which is initially vaguely expressed as t1, can also be expressed as t2". In the course of this process it may happen that one same term turns out to have different and opposite meanings. When this happens, a first modification of the meaning of t1 emerges: t1 appears as the universal concept ("being") and as the universal concept insofar as it is determined as t2, through exclusion of the opposite of t2 ("matter as continuous, and not discrete" or "being as being, and not not-being"). The second step implies that the two antonymic determinations appear as reciprocally implying each other, the one cannot be determined without the other, hence the universal t1 is now determined as both t2 and not-t2; the third final modification of the meaning of t1 consists in seeing the antonymic determinations as forming together the subject of discourse. As such they loose their antonymic character. Differently from Fulda, I claim that the difference between the second and the third dialectical step do not imply that the opposites loose their antonymic character. Rather, I follow Michelet's distinction between "contradicting oneself" and "seeing the contradiction" in Michelet 1871, 24–41 applying it respectively to the second and the third dialectical step. In the second step each opposite implies its negation (i.e. it contradicts itself), in the third we gain a level in which we see that the opposites imply each other (in which we see the contradiction – I examine what this may mean in logical terms in Part V). On Fulda's analytical interpretation of Hegel's dialectic Düsing 1976, 315 writes "[For Fulda] the beginning of Hegel's logic is the vagueness [of natural language] which, through progressive modifications of meaning, assumes a more precise meaning. In this [reconstruction], two aspects are inevitably omitted, namely Hegel's view about the self-movement of the concept as the concrete universal, and the necessity of the logical contradiction emerging within the dialectical process". Also Marconi 1979b, 23 stresses that in Fulda's

necessity". During a conference on formal logic and dialectic,[35] Hintikka also remarked:

> [Not only] have we not [...] managed to address ourselves to the theme of this meeting, i.e. to compare with each other formal logic and dialectic. It seems to me that we have failed even more radically. We have not really located any unified phenomenon which would be termed 'dialectic' and which could be compared with formal logic. I have tried to keep a list, as it were a score sheet, of the different senses of dialectic which have made their appearance in [our] discussions [...] When the count recently exceeded 20, I began to worry. (Hintikka 1981, 110)

In this context, I hold that some logical peculiarities emerging from Hegel's theory of dialectical inferences in the *Lectures on the History of Philosophy* are worthy of being highlighted: they have not been considered in the literature, and yet they are essential. They show dialectics' clear link to the history of logic, and

---

account, contradictions are a casual, and not necessary, product of linguistic analysis. Salermijn 1971 and Butler 1975, 414–431 claim that Hegel deals with contradictions in orthodox terms, i.e. adopting the method of *reductio ad absurdum*. This approach is problematic, in particular if one recalls Hegel's explicit distinction between *reductio ad absurdum* arguments and genuine dialectics in the *Lectures on the History of Philosophy*. Marconi (1979b, 26) explains that while in a *reductio* argument the premise that leads to the contradiction is negated, "in Hegel's philosophical discourse the 'premise' of the contradiction is indeed rejected as adequate expression of the absolute, but it is also maintained – together with its contradictory consequences, as partial truth and as adequate expression of that particular moment of the development of the absolute". This means for Marconi that "from an orthodox point of view a premise that is refuted through deriving from it a contradiction is – so to say – abandoned once and for all: it will not belong at all to the true theory. Not so in Hegel: the premise and its contradictory consequences fully belong to the true theory, precisely because the true theory is identified with all the steps needed to argue for its truth". Importantly, Marconi stresses that the premises in a dialectical inference have partial truth, and that the full truth is the whole theory. More precisely, according to Marconi 1979b, 43 dialectical inferences involve three aspects. First, the indeterminacy of natural language, in dialectics, always generates contradictions, and second the discovery of contradictions forces us to abandon the claim that our conceptual determinations are adequate to express the conceptual content, and forces us to look for further determinations. Finally, the conclusion in dialectical inferences is not contradictory. Differently from Marconi, I claim that the conclusion of a dialectical inference and the adequate form of the conceptual content is the (biconditional) contradiction. Routley/Priest 1984, Chapter 2, and more recently Bordignon 2014, 78 also stress the closeness between paradoxical deductions and dialectical inferences. One difference between these accounts and mine is that, in my view, the more adequate logical characterisation of the conclusion of a dialectical process is the biconditional contradiction, and not the conjunctive contradiction.

35 The proceedings appeared in 1981 with the title *Konzepte der Dialektik*. See Becker/Essler 1981.

are helpful for assessing the question about Hegel's general view on validity. In a sense, my attempt can be seen as a contribution to increasing the confusion lamented by Hintikka. But perhaps locating Hegel's view within the history of logic is needed to settle the matter once and for all concerning the question: "What is Hegel's dialectic?".[36]

12.1 Hegel's account within the history of logic

In the overview on Hegel's *Lectures on the History of Philosophy* I have already tried to emphasize some recurrent mechanisms at work in dialectical inferences. Hegel uses some significant metaphors and circumlocutions to describe them:
- The "marvellous fact"
- The "inconsequence" resulting from the "extreme consequence"
- The "scratch of limitation"
- The "positive result"

---

[36] The question about the logical meaning of dialectical inferences has gained a new attention in the last 20 years, also thanks to the attempts at reading Hegel in analytical perspective (see among others Horstmann 1984, Stekeler-Weithofer 1992, Brandom 2002, McDowell 1994, Ruggiu/Testa (eds.) 2003). See Bencivenga 2000, d'Agostini 2000, Redding 2007. Berto 2005 and 2007b, 19–39 has given a detailed account of the meaning of dialectical inferences based on Brandom's inferentialistic reading of Hegel. Berto 2007b, 19 writes: "My reading is based on a very simple idea: the inferential intuition that an essential part of what it is to grasp a conceptual content, and to be able to apply it correctly to an object, consists in mastering its connections with the concepts it entails, and with the concepts that entail it [...] Brandom ascribes such an inferential holism not only to Sellars but also to Hegel. [My work] aims at showing why this ascription is quite correct". For Nuzzo 2010b, 61–82 Hegel's logic should be read as a program of clarification and revision of language – both of ordinary language and of the language of traditional logic and metaphysics. That dialectics develops as revision and articulation of the meanings of linguistic terms is what reveals the historical component in Hegel's conception of thought and language. Nuzzo also highlights one aspect emerging from Hegel's review of Göschel's Aphorism, namely the question about the difference between representative and conceptual language. As Nuzzo 2010b, 65 stresses, Hegel distinguishes between the two kinds of language and states that "all scientific mediation consists in the double movement of crossing over from representation to concept and from concept to representation" (see Hegel Werke 11, 378). Hence dialectics implies a double movement from representation to concept and vice versa (whereby the language of representation cannot but be dogmatic or fixing, while the conceptual one is sceptical and negative). Bordignon 2013, 179–198 discusses the interpretation of dialectics as analysis of the vagueness of natural language as given by Marconi 1979b and Nuzzo 2010b. In 2013, 35–52 I compare dialectical logic with dialetheism, the perspective according to which there are true contradictions, examining Priest's own interpretation of dialectics as dialetheism in Priest 1989, 388–415. Bordignon 2014 presents a dialetheist reading of Hegel's dialectics.

These are indeed just metaphors of logical relations, but they express some aspects of dialectical validity that are especially relevant from a logical point of view.

The "marvellous fact". Between consequentia mirabilis-arguments and paradoxical deductions

As we have seen, Hegel explains that a "marvellous fact" meets us "when we take determinations such as these", i.e. the fundamental concepts of reason such as "being", "the one" etc., "by themselves" (Hegel Werke 19, 80/Hegel 1892ff., vol. II., 57–58). "Taking determinations in themselves" or, which is the same, considering concepts purely, involves asking the Socratic questions: "what is being?" "what is the one?" "is the one one?" etc., to reflect, in other words, on our own conceptual practices.[37]

What is "marvellous" ("*das Wunderbare*") for Hegel is the fact that, when we take concepts *by themselves*, trying to analyse them, the statements about them turn into their opposites, i.e. imply their negation.

Hegel's terminology and the described "fact" recall what in the history of logic is known as *consequentia mirabilis* (CM), an inferential form which permeates the whole history of philosophy, and is strictly related to reasoning about truth.[38] A short comparison may shed light on some aspects of the Hegelian point of view. The form is usually (Bellissima/Pagli 1996, 7) defined as follows:

> CM: *If from the negation of a sentence A we deduce A, then A is true*

---

[37] That dialectical inferences involve, in this sense, semantic ascent has been stressed by many authors, and is a characteristic insight of non-classical readings of Hegel's dialectic. For Marconi 1979b, 19ff. dialectic requires the application of terms to themselves, and thus that the usual restrictions of type do not work. The violation of the restrictions of type on the basis of classical logic (in which the classical rules hold) generates the emergence of antinomies. Findlay 1981, 132 observes that dialectical inferences imply what he calls a *metabasis*, that is: "that genuine passage beyond premises that is also involved in passing from an object language to a meta-language [...] and in which [a conclusion is implied] by its premises rather in the sense in which G. E. Moore said that to assert that it is raining is to imply that one believes that it is raining". Kulenkampff 1970 claims, similarly, that dialectic implies the attempt at expressing the semantic properties of the language within the language (in Tarksian terms, this means that the language contains its own truth-value).

[38] For the history of *consequentia mirabilis* see Bellissima/Pagli 1996. An analysis of the philosophical meaning of CM-arguments can be found in d'Agostini 2002, Id. 2009, Chapter 10 and Id. 2011.

Early non-technical formulations of the form can be found in Democritus, Plato, Aristotle, Euclides, Anselm, Thomas Aquinas, and Burleigh. The concept and the term entered the logical discussion starting from Gerolamo Cardano in 1570, and Clavius (the law is also known as Clavius' Law).[39] In Plato's *Theatetus* Socrates refers to a "highly beautiful inference" concerning Protagoras's thesis "everything is true". The thesis implies that what Protagora's opponents say, namely that Protagoras is wrong, is true too, hence "Protagoras will concede that his own view is false, if he thinks that what his opponents say is true" (Teetetus 170, XXII, 219–220 – translation from Hamilton/Cairns 1961).

Similarly, Aristotle argues in the *Metaphysics* that "all these doctrines (i.e. views according to which everything is true, or that everything is false) fall into the inconvenient of destroying themselves" (Aristotle *Metaphysics* Gamma IV, 8 1012 b 12–22 – translation from Barnes 1984). Those who state that everything is true consider true the negation of their own view, and from this it follows that their own view is not true; those who hold that everything is false state that also the thesis they are stating is false, and thus that it is not the case that everything is false. Anselm uses it in order to demonstrate the eternity of truth. Thomas Aquinas generalises this result, stating that "every entity whose existence implies the destruction of its own existence is eternal".[40]

In modern logic the rule is explicitly accepted by Russell, Whitehead, Frege. Its philosophical implications are discussed in C. I. Lewis. Russell (1906, 159–202) formalises it as follows:

CM1: $(A \rightarrow \neg A) \rightarrow \neg A$

and its correspondent positive form as

CM2: $(\neg A \rightarrow A) \rightarrow A$

and considers both CM1 and CM2 as specific declinations of *reductio* arguments, whose form is

reductio: $A \rightarrow (B \wedge \neg B) \vdash \neg A$

---

[39] See Bellissima/Pagli 1996, 11–12.
[40] See Bellissima/Pagli 1996, 202.

In both CM- and *reductio*-argument, the conclusion is the negation of the original assumption. However, that CM-arguments are specific versions of reductio arguments is arguable.[41] One first evident peculiarity of CM-arguments with respect to *reductio* arguments is the immediacy of the passage.[42] That is, the fact that we draw ¬A immediately from A, without intermediate passages. Not only that, CM arguments are classically traced back to arguments aiming at rejecting scepticism.[43] This is the aspect that is perhaps most interesting from the Hegelian perspective. I will get back to it in the next section. For now I will focus on the differences between the derivation of a single sentence from its negation in CM arguments and dialectical inferences.

As a matter of fact, Hegel's "marvellous fact" can be taken to belong to the history of and discussions on *consequentia mirabilis*. The evident continuity is the fact that in both CM arguments and dialectical inferences *from the negation of a [single] sentence A we deduce A*. However, *while this deduction in CM arguments means that A is true, in dialectical inferences the same deduction means that A is not true*. We can recall Plato's description of the dialectical practice considered by Hegel in his *Lectures* as an example of genuine dialectic:

> [T]he many turns into the one insofar as it is considered in the determination it has. This is the marvellous fact that meets us in thought when we take determinations such as these by themselves, is that each one is turned into the reversal of itself [...] In the sentence "the one is [one]" is implied "the one is not one, it is many" and conversely "the many is [many]" simultaneously implies "the many is not many, it is one". They manifest themselves as dialectical, they are, essentially, the identity with their negations; and this is their truth. (Hegel Werke 19, 82/Hegel 1892ff., vol. II., 59f.)

In sum, the Hegelian "marvellous fact" does not allow us to conclude, as we are supposed to do in a CM-argument, ¬A from A → ¬A, or A from ¬A → A. In fact, the same "marvellous event" that takes place when we apply semantic ascent to A, the "event" of A turning into ¬A, happens when we apply semantic ascent to ¬A, asking "is ¬A true?" or "is it true that being is not not-being?". Thus the Pla-

---

[41] On the differences (as well as on the debate on the differences) between CM and *reductio* arguments see Bellissima/Pagli 1996, 153.
[42] d'Agostini 2009, Chapter 3 distinguishes between two kinds of reductio and consequently two kinds of paradoxes depending on the respective failures of the two reductions: "We have two different cases of *reductio ad absurdum*. The first is reductio per self-refutation: I eliminate the thesis a because from a it follows $\neg a$ (given $a \rightarrow \neg a$ I assume $\neg a$). The second is normal reductio: the premise a generates a contradiction, and thus it is to be negated (given $a \rightarrow (b \land \neg b)$ I assume $\neg a$). The paradoxical situation arises because reductio fails.
[43] See Bellissima/Pagli 1996, 127.

tonic argument shows that we do (necessarily, deductively) infer from A (the one is one) ¬A (the one is not one) but that, since we also necessarily infer from ¬A that A, all we have is the biconditional $A \leftrightarrow \neg A$. Analogously, in Hegel's analysis of the Liar paradox the same inference form (from A: "s/he lies" to ¬A: "s/he does not lie" and vice versa) typical of Plato's dialectic (from A: "the one is one" we infer ¬A: "the one is not one" and vice versa), is at work. Moreover, as we have seen Hegel claims that the two answers "A" and "¬A", taken in isolation, are wrong ("a simple answer cannot be given").

In this respect, the structure of dialectical inferences, though sharing features with CM-arguments, corresponds rather to paradoxical deductions.[44] The relation between CM-arguments and paradoxes is implicitly acknowledged in contemporary discussions.[45] The Liar paradox, the sentence L that says of itself that it is false, releases the two arguments $L \rightarrow \neg L$ and $\neg L \rightarrow L$.[46] As d'Agostini notes:

> [T]he two arguments correspond to a self-refutation [...] (from L we derive that non-L) and a self-foundation (from non-L we derive L). We can thus say that an antinomy combines a self-refutation and a self-foundation, or also that self-refutations and self-foundations are "half-Liars". (d'Agostini 2009, 140 f.)

In this respect we can highlight the specific nature of dialectical deductions also with respect to this account of paradoxical deduction as a combination of self-

---

[44] The formal similarity between dialectical inferences and paradoxical deductions is stressed by Priest/Routley 1984, 92–93. I agree with Priest/Routley 1984 about the formal closeness between dialectical and paradoxical inferences: both are sound arguments conveying a true contradiction. However, my claim is that Hegel has a different view of truth and its link to contradiction than the one endorsed by Priest and Routley in 1984.
[45] See for example Sainsbury 2009. The connection is explicitly addressed and explained by d'Agostini 2002, 2008a and 2009, Chapters 10 and 11. It is also hinted at by Bellissima/Pagli 1996, 151 ff.
[46] A possible way to express the Liar paradox (L) formally is:
(L)
L: L is false (this is the formal expression of the Liar sentence "this sentence is false")
The Liar sentence immediately implies:
$L \leftrightarrow \neg T``L"$ (where "T" is the truth predicate, and falsity is the negation of truth)
If we assume the truth-schema:
$Ta \leftrightarrow a$
And apply it to L we have
$T``L" \leftrightarrow \neg T``L"$. The reconstruction is taken from Cook 2009, 171. The passage from L: L is false [formally expressed as: (L) "FL"] to $L \leftrightarrow \neg L$ is usually explained via the capture- and release-behaviour of the truth predicate in the T-schema (and the definition of falsity).

foundations and self-refutations. In other words, a self-refutation establishes a *necessarily false* thesis, while a self-foundation establishes a *necessarily true* one. They can be expressed as $(A \to \neg A) \to \neg A$ and $(\neg A \to A) \to A$ which, as we have seen, correspond to the two "formulations, negative and positive, respectively, of the rule called *consequentia mirabilis*" (d'Agostini 2009, 113–114).

In Hegel, the fact that from $\neg A$ we derive A does not mean that A is true, and, vice versa, for Hegel from the deduction of $\neg A$ from A we cannot infer that $\neg A$ is true. Hence dialectical inferences are to be distinguished from both CM-arguments and paradoxical deductions. What Hegel underlines is that the inferential *movement* from A to $\neg A$ and back is true – what is true is the biconditional $A \leftrightarrow \neg A$, whereby A and $\neg A$ are untrue. "The nature of the concept is so powerful" Hegel writes in a book review from 1929 "that in an untrue sentence is already entailed (often already explicitly stated) the opposite determination to the one which is asserted" (Hegel Werke 11, 380). That we derive $\neg A$ from A means that A is untrue; that we, in turn, infer A from $\neg A$ means that also $\neg A$ is untrue.[47] As we have seen, Hegel also states that being and nothing, which "pass into each other" are not the true concept. The true concept, the truth about being and nothing, is becoming, which is the movement between the two, their "unity", and which is distinct from them. The "movement" and "unity" are, evidently, "figures" of the biconditional.[48]

The positive addition, the scratch of limitation and CM again

In their *The Development of Logic* Kneale and Kneale write, as we have seen, that "[In Plato, dialectic is] the hypothetical method of refutation together with some mysterious positive addition" (Kneale/Kneale 1962, 10). From the point of view of Hegel's history of dialectic the "positive result" of dialectical refutations cannot be equated, as Kneale and Kneale do, to "Zeno's method of refutation".

Priest and Routley 1984 show that what the Kneales call "mysterious" is not difficult to understand, but rather "quite straightforward. The Kneales run into trouble through presupposing a dubious positive/negative distinction, linking *reductio* arguments and refutations as negative invariably with negative results. But Zeno's procedure already indicates how results such as Parmenides' thesis that motion is impossible, a thesis of high generality, can be enforced by dialec-

---

[47] I will examine the consequences of these ideas for the meaning of negation in the last part.
[48] In this respect, dialectical inferences are also different from paradoxical deductions as presented and explained by Priest/Routley 1984. I will address this difference in Part V.

tical methods, e.g. supposing the opposite and deriving unacceptable conclusions" (Priest/Routley 1984, 86 footnote 21). Priest and Routley stress here that the supposedly mysterious positive addition (conveyed by Platonic dialectic) is simply the function that refutations display in enforcing theses of high generality.

I suppose instead that what Kneale and Kneale refer to is something else. More precisely, the question at stake is the specific difference between Zeno's *reductio* and Plato's dialectical method, a difference that clearly emerges in Hegel's account. The "positive result" of genuine, Platonic dialectic is at the very core of Hegel's discussion, and one could argue therefore that Hegel does contribute to solve the mystery.

As we have seen, Hegel explicitly distinguishes Zeno's arguments, called "immanent dialectic with negative result", from Plato's genuine dialectic, called "immanent dialectic with positive result". Similarly, he distinguishes genuine positive dialectic from other "negative" dialectics, such as the ones performed by the Megarians and by the Sceptics.

As we have seen as well, Hegel recalls as an example of genuine dialectic the controversial passage of Plato's *Sophist*. According to Hegel's interpretation the passage shows that what Plato had in mind is very different from the situation in which a highly general thesis (such as Parmenides' view about the impossibility of motion) is *enforced* by reductio-arguments. We have rather what Hegel calls "true contemplation" (*wahre Einsicht*), translating the Platonic word *elenchos*, and namely the fact that from a thesis of high generality (let us call it A) we infer its negation (not-A), and from its negation (not-A) we are forced to infer A. In Hegel's terms, the Platonic positive addition is the insight into the fact that the resulting contradiction "does not disintegrate the concept" but is rather "its truth".[49]

In this respect, d'Agostini's explanation of the difference between dialectics and Sophistic is useful to understand the Platonic (and also Hegelian) point:

> Plato's and Socrates' dialectics is identical to the one of the Sophists, but interested in the logical behaviour of words, in particular of some words: truth, the good, justice, existence.

---

[49] Sarlemijn 1971 and Butler 1975 interpret Hegel's dialectic as involving standard reductio-procedures. On the difference between dialectical arguments and reductio-arguments Marconi 1979b, 26 writes: "In Hegel's philosophical discourse the 'premise' of the contradiction is indeed negated as adequate expression of the absolute, but it is also maintained – together with its contradictory consequences – as partial truth [...] from an orthodox point of view a premise which is refuted through deriving from it a contradiction is – so to say – abandoned once and for all: it will not belong to the true theory, to which belongs its negation. Not so in Hegel: the 'premise' and its contradictory consequences fully belong to the true theory".

> These concepts escape the sophistic nullification, they are *an-elenchtic*, insofar as they bring about irrefutable theses. (d'Agostini 2011a, 265)

As d'Agostini stresses, the Platonic "genuine *elenchos*" is strictly linked to the irrefutability (*an-elenchtic* nature) of some concepts. This confirms the Hegelian discussion about the failure of scepticism if applied to rational concepts (such as truth, being, the one etc.). While the single theses of a dialectical contradiction involve their negation (they refute themselves) the contradiction itself, which expresses the whole nature of the concept, is true, that is, irrefutable.

Since, as Hegel says, the true meaning of the concept of being is formally expressible by the (inferential) movement from one determination (p: "being is nothing") to its negation ($\neg p$: "it is not the case that being is nothing") and vice versa, that is by the biconditional: $p \leftrightarrow \neg p$, then the sceptical refutation – consisting in stating "for every logos there is an opposite one which is equally valid" fails. In the example taken from *Adversos Mathematicos* Sextus takes the concept of reason, which is a contradictory concept, as having either one or the other of two contradictorily opposite properties ("being the whole" and "being a part", which, for Hegel, are formally expressible as one the negation of the other). He separates the antecedent and the consequent of the biconditional expressing the whole concept of reason: "$p \leftrightarrow \neg p$"), stating p, and derives its negation; and then he takes $\neg p$, and derives its negation. In so doing Sextus does not "scratch", i.e. confute, reason, but merely develops its formal structure. In separating the antecedent from the consequent of the biconditional he "gives to reason the scratch of limitation", which means that he *misunderstands* reason – he transforms or alters what is complex and heterogeneous (contradictory) reducing it to something univocal.

Thus, the sceptic tries to refute rational concepts, concepts whose structure is the contradiction $A \leftrightarrow \neg A$. Hegel shows that in trying to refute reason, the sceptic affirms reason itself. Hence we admit, again, a CM-argument, according to which from the confutation of reason, which is a contradictory content, we derive reason itself. The argument could be taken to have the following form:

$$\neg A \rightarrow A \vdash A \leftrightarrow \neg A$$

or also:

$$A \rightarrow \neg A \vdash A \leftrightarrow \neg A$$

Dialectical concepts are truly contradictory concepts, if we try to confute them they appear again, escaping the sceptical (and sophistic) nullification.

12.2 Controversies on the nature of dialectical inferences

Interpreters do not agree on the meaning of validity at the basis of dialectical logic. The controversial points are the binomials syntactic versus semantic, deductive versus inductive, and material versus formal validity.

In the first attempts at formalising dialectic from a non-standard logical point of view, as documented in Marconi 1979a, it is already clear that dialectic involves an interplay between semantics as theory about both the meaning of concepts and of truth, and syntax as theory about the form of our knowledge about them. In both Apostel 1979 and Routley/Meyer 1979 it is evident that dialectic implies the view of logic as *interplay* between syntax (form) and semantics. In 1979, Routley and Meyer stress that dialectic has a semantic relevance. To this effect they quote Osnovy who in 1958 remarks "the fundamental question of dialectical logic is the problem of truth. Dialectic examines the forms of thought with respect to the content and shows to what extent they mediate a true knowledge of the world" (Osnovy 1958, 327 quoted by Routley/Meyer 1979, 328). That the fundamental question of dialectical logic is the problem of truth means, in other words, that dialectical is the consideration of forms insofar as they are able to convey true content, and not that dialectic has a semantic and thus non syntactic nature. From this perspective the notion of dialectical validity reveals itself to be perfectly compatible with the contemporary notion of semantic validity.

Another typical controversy concerns the question whether dialectical inferences are deductive or inductive. Findlay denies that the propositions in which Hegel develops his *Science of Logic* are connected deductively. According to Findlay (1958, 149–150) Hegel's dialectics "has little in common with a deductive chain of propositions" and the validity of his system is not "that of a proof or a proven theorem". The *Science of Logic* for Findlay is rather a chain of linguistic recommendations:

> Hegel recommends for our adoption of a given way of talking about the world, then discovers flaws and inadequacies in this mode of speaking, then supersedes it by a further recommendation which also comprehends it, until his last recommendation supersedes and comprehends all others [...] It is obvious that there can be no question (in the ordinary sense of the words) of either truth or validity in such a series of recommendations. There can only be questions regarding the linguistic or conceptual adequacy or satisfactoriness of its terms. (Findlay 1958, 151)

Kulenkampff (1981, 140–144) claims that dialectical validity is neither inductive nor deductive. He hints at the continuity between Schelling's and Hegel's view on the matter, writing that for Schelling dialectic is a special sort of inductive reasoning, which derives "the elements from thought [and not from experience]". Through it "philosophy [which is a searching discipline] arrives at the principle".

Other authors accept that Hegel did want to develop dialectic scientifically, i.e. deductively, but stress that he failed. Findlay (1955, 14) himself says that "Hegel never achieves deductive rigour in his practice: it is only his *account* [my emphasis] of that practice which suggests that he is aiming at it". Findlay (1955, 15) recalls Croce who "has both recognized and admired the poetic character of Hegel's dialectical proceedings". Similarly, Gadamer 1976 stresses that, even if Hegel complained about the unscientific way in which dialectic was practiced by his contemporaries and tried to develop dialectic in deductive form, his attempt at developing the inferential movement of rational thought in a scientific way, as fixable and repeatable logical succession of consequence-relations, failed. For Butler (2012, 13), on the other hand, "dialectics is deductive only up to the point where it implies a dialogue between different voices in rupture with one another. It is deductive only insofar as it carries out the logical implications contained in a single voice".

The historical account that I have proposed, whose aim was in locating Hegel's view within the history of logic, may provide suggestions for resolving these controversies.

Though claiming that the rational (*vernunftlogisch*) way of thinking seems inconsequent if compared to the standard logical, intellectual one (*verstandeslogisch*), dialectical logic stands in fundamental continuity with the common contemporary notion of semantic validity as truth preservation in virtue of form. To be more precise, dialectical inferences could be defined as a particular declination of semantically valid inferences, namely as actually truth-conveying inferences, whereby contradiction is seen as the norm of truth.[50] Hegel identifies the specifically "dialectical consequentiality" with the one that manifests itself in nature, or juridical institutions:

> The tree has at first no flowers or leaves, it seems to be dead, then it starts to have leaves, flowers and branches, which then change, the leaves and flowers fade, and fruits and seeds take their place. Thus this change, the dialectic, breaks boundaries, it becomes inconsequence. (Hegel 1992, 13)

---

[50] I agree with Schäfer 2013, 265, who stresses that to think logically, for Hegel, is to think in terms of contradictions, i.e. consists in developing and overcoming (in the sense of *aufheben*) the contradiction. I highlight that this view does not imply a rejection of the traditional notion of valid inference. Rather, it links validity to a philosophically and speculatively understood truth.

And:

> The law [...] is held within boundaries by the understanding. Amnesties and absolutions are something beyond the boundaries of the law, the law as limitation finishes here, and the inconsequence emerges. (Hegel 1992, 13)

Plants are seemingly dead and then live again, flowers suddenly disappear and are replaced by fruits, people are sent to prison, and then (occasionally) absolved. All this is indeed a sign of deep inconsequence, and yet it happens, and we do not find it absurd or wrong.

What Hegel hints at here is the peculiar nature of dialectical inferences that we have seen in considering the *Lectures on the History of Philosophy*. The dialectical passage from premises to conclusions is equated to the plant's dying and then living again, or to the unexpected release of guilty people during amnesties. Again we see here the relation, highlighted above, between opposite elements/views/conceptual determinations (being alive, being dead; being captured, being released), which are united in a more general, comprehensive concept (the natural process; the law). The same happens, as we have seen, with the conceptual determinations examined in Hegel's logic. Thus we see that the dialectical inference relation implies a necessary, and yet "surprising" connection between premises and conclusions.

In other words, dialectical inferences, as we have seen, do not consist in stating "p" ("justice is the advantage of the stronger"), adducing grounds for it ("because justice is expressed by the laws and the laws are made by the stronger and in the interest of the stronger") and dismissing counter-arguments. Rather, when we think dialectically we think philosophically, that is, as we have seen: we are interested in the analysis of the concepts involved in our assumptions (the concept of justice, the concept of strength etc.), and we do not assume anything as given and true once and for all, but question the truth and validity of our assumptions and arguments. As Hegel claims: "what is taken for truth [...] can only deserve the name of 'truth' when philosophy has had a hand in its production" (Hegel Werke 3, 63/Hegel 1977, 41).

In the *Lectures on the History of Philosophy* this peculiar, philosophical nature of dialectical inferences clearly emerges. I have hinted at its formal nature in discussing the relationship of dialectical inferences to CM-arguments and paradoxical deductions. According to my reading, that dialectical arguments are seemingly "inconsequent" simply means that they involve inferences such as $A \rightarrow \neg A$ and $\neg A \rightarrow A$, from which we logically derive: $A \leftrightarrow \neg A$.

However, that dialectical inferences necessarily convey contradictions does not mean that Hegel rejects the standard view of logical consequence as truth preservation in virtue of form.

12.3 Truth and Validity

In contemporary logic it is in use[51] to distinguish between two notions of validity, syntactic and semantic validity. Only in the latter truth, i.e. a certain link (the classical Aristotelian one) between "how things are" and inferences plays a decisive role. The two notions are of relevance in relation to Hegel's perspective. According to the concept of syntactic validity, an argument is valid if it follows the logical rules of the language in which it is formulated.[52] In this sense, the necessity of the logical consequence is basically linguistic. The normativity of logic is based on the logical constraints produced by logical languages, or systems. In traditional logic, logical constraints (rules) come from thought (are rules of thought).

According to the notion of semantic validity, validity is based on truth. In the semantic definition, an inference is valid if, given the truth of the premises, the conclusion is also true. This is the so-called requisite of truth preserving. Here the logical constraints come from the world, or from the facts that make true the premises and/or the conclusion. But notably, the formula "if the premises are true, the conclusion is also true" does not say anything about the real effective truth, truth with reference to the world in which we live. The expression "given the truth of the premises" means "in every possible world in which the premises are true". The truth presupposed is possible or hypothetical.

The two notions are not rival. They simply come from looking at validity from different perspectives. However, it should be noted that the first notion of validity is established by the axioms or rules of language, so – as it seems – is ontologically neutral; while the second is ontologically committed, with reference to some world. Yet the world at stake is a possible world, or rather: a set of possible

---

[51] See for the following discussion Haack 2006, 13ff., Read 1995, 35ff., Priest 2006, 176ff. For the distinction between proof-theoretic and model-theoretic validity see Beall/Restall 2014.

[52] According to Haack a sequence of sentences of a formal language L (the premises of the argument): $A_1 ... A_{n-1}$ is syntactically valid in L just in case A_n (the conclusion) "is derivable from $A_1 ... A_{n-1}$, and the axioms of L, if any, by the rules of inference of L", Haack 2006, 13. Priest states that in the proof-theoretic (i.e. the syntactic) account "one specifies some basic rules of inference syntactically. A valid inference is then one that can be obtained by chaining together, in some syntactically characterizable fashion, any of the basic rules" (Priest 2006, 177).

worlds. So we can say that in both cases validity is a formal phenomenon, and the enterprise of logic is still to establish validity "in virtue of form".[53]

The semantic account of validity is also read in model-theoretic terms. In the proof-theoretic account of logical consequence validity amounts to "there being a proof of the conclusions from the premises",[54] in the model-theoretic one validity is based on the notion of truth in a model. Models are structures that provide possible interpretations for each term in the formal language. On this account I can show that an argument is valid iff in any model in which the premises are true the conclusion is true too. In contemporary discussions[55] the proof-theoretic account is said to be more compatible to an anti-realist, the model-theoretic one to a realist approach in philosophical logic. In the former, inference rules are usually taken to be basic – as definitions of the meaning of the operators. In the latter, "models" are identified with worlds, and the analysis of "truth in models" is interpreted in terms of explication of truth as correspondence to reality or the world.[56]

All this stated, the notion of semantic validity implies that the premises do not need to be true at the actual world in order for the conclusion to be true. We can have perfectly valid inferences without them saying anything about the actual world. As we have seen, in discussing Aristotle's syllogistic Hegel underlines that

> [T]he form of an inference may be absolutely correct, and yet the conclusion arrived at may be untrue [and this is the sign that] this form as such has no truth of its own. But from this point of view these forms have never been considered. (Hegel Werke 19, 240/Hegel 1892ff., vol. II., 222f.)

---

[53] See Sider 2010, 2 and Read 1995. As Read observes: "an argument-form is valid if, however the schematic letters are interpreted, the result does not consist of a collection of true premises and a false conclusion [...] logical consequence is a matter of form, namely, that however the schematic letters are interpreted, truth is preserved from premises to conclusion: we never obtain true premises and false conclusion" (Read 1995, 38).
[54] Beall/Restall 2014.
[55] See the reconstruction in Beall/Restall 2014.
[56] Yet the view that a model theoretic approach is realistic can be questioned. Berto 2005, 180 criticises the *ad-hocness* of the model theoretic account: "individuating a 'semantics' for a calculus intended as a certain kind of model with respect to which we can prove the consistency and completeness of a system is not too difficult, if we rely on algebraic structures coined *ad hoc* [...] From an interesting semantics we should demand more: to be describable independently from the resources given by the same language on which the formal system is based".

That classically valid inferential forms often convey absurd or plainly false conclusions, and are not able to give an account of our scientific (and philosophical) inquires, is stressed today in non-classical logics, such as intuitionistic, modal, relevant logics. The need for criticising classical inferential forms, finding new ones, is the very reason for the birth of non-classical logics.

I have already stressed (see chapter 6.) that Hegel's critique of *Verstandeslogik* anticipates many points highlighted in non-classical logics. What I wish to show now is that Hegel's critique of the validity of some inferential forms does not imply a dismissal of the general idea of semantic validity, or a de-legitimation of the formal logical enterprise in favour of informal accounts of reasoning.

The Aristotelian notion of logical consequence, according to which

> certain things having been supposed, something different from those supposed results of necessity because of their being so

is at the basis of the Hegelian one. In the quote above, Hegel criticises *Verstandeslogik* by arguing that its forms of valid inference are not able to convey truth. The reason is that the logicians of his times simply assume that the forms they fix are good (i.e. forms of truth). But they are not, according to Hegel, and this is the reason why Hegel's logic enfold as a critical analysis of the forms fixed by *Verstandeslogik* specifically in view of the question: "are they truly forms of truth?".

According to Hegel the question about the truth of the forms (about their ability to express how things stand) is fundamental. Truth is the very condition of the validity of the forms. As we have seen, the "truth" logic deals with for Hegel (*Wahrheit*) is conceptual truth, and not the contingent and contextual coincidence of our thoughts with reality. "Conceptual" does not mean product of conventions, or psychological or mental. Conceptual thought is complete and sceptical thought as the condition of our thought's correspondence with how things really stand.

Hence Hegel's dialectical logic can be understood as the analysis of conceptual truth as the condition of inferential validity. In other words: for Hegel the Aristotelian idea of "having supposed certain things others follow with necessity" is perfectly in order. Dialectical logic deals with the very condition or reason of the necessity of the passage from premises to conclusion in arguments, namely truth. Truth, as we have seen in Part III, is for Hegel thought's correspondence with reality, whereby the kind of thought actually able to express reality is rational or speculative, i.e. complete and sceptical, thought. Conceptual thought, conceptual analysis, complete and sceptical determination of the meaning of the conceptual words is the basis of genuinely valid and sound inferences. This is not an extra-logical or non-formal requisite because Hegel does individuate in the contradiction A $\leftrightarrow \neg$A the form of this kind of thought, i.e. of truth.

Summary

The fourth part is focused on a concept – the concept of validity – which, according to contemporary logicians, is the fundamental logical concept. Mignucci declares Aristotle founder of the discipline we now call "logic" on the ground that Aristotle, for the first time, develops a systematic distinction between valid and invalid inferences. Yet the concept of validity is a relatively new subject of study – in Hegel the term (which in German is sometimes translated as *Folgerichtigkeit*) does not figure in a technical meaning. The notion of validity at the basis of Hegel's logic is the one fixed by Aristotle in the *Prior Analytics:* [in a logically valid inference: *syllogismos*], certain things having been supposed, something different from those supposed results of necessity because of their being so.

In an important passage on Aristotle in the *Lectures on the History of Philosophy* Hegel writes that "the form of an inference may be absolutely valid [*richtig*], and yet the conclusion arrived at may be untrue [and this is the sign that] this form as such has no truth of its own. But from this point of view these forms have never been considered" (Hegel Werke 19, 240). Hegel's logical theory has the aim to empower the very notion of validity fixed by Aristotle, grounding it in the notion of (a speculatively/dialectically understood) truth [*Wahrheit*]. This is the reason why this part enfolds as an analysis of Hegel's notion of dialectical inferences (the inferences that, for Hegel, are the genuinely valid ones), and his distinction between valid (genuinely dialectical) and invalid (non dialectical) inferences.

In **Chapter 10** I examine Hegel's definitions of dialectic and his analysis of dialectical inferences from a historical point of view, i.e. presenting Hegel's interpretation of dialectic from Zeno to Kant in the *Lectures on the History of Philosophy*. I highlight how three aspects that are canonically associated with the concept of dialectic are united in Hegel's view: dialectic as movement of pure (i.e. reflexive, self-referential) concepts; dialectic as the art of the dialogue; dialectic as the logic of contradictions. In the *Lectures* Hegel distinguishes between several declinations of dialectic: internal and external, internal with positive result and internal with negative result. Schematically, "external dialectic" is the technique employed by the sophists aiming at producing contradictions in every discourse (arguing for example that honey is sweet and that it is not sweet, or that six is great and six is small), "external" means disconnected from the task of analysing conceptual determinations, having as sole aim the defence of particular interests. Internal dialectic is in contrast linked to the aim of analysing conceptual determinations, such as "being", "the one", "movement". Zeno is the discoverer of *internal dialectic* (the Megarians, the sceptics, and Kant develop different kinds of this same internal dialectic with negative result). He is interested in determining the meaning of motion, he analyses the concept of motion and, in so doing, discovers necessary contradictions within the concept. Zeno's dialectic is, however, not yet the genuine one: it has negative result because, for Zeno, the emergence of necessary contradictions disintegrates the concept: the assumption "motion exists" generates contradictions, and this means that it is false. Hence Zeno's *internal dialectic with negative result* is simply the method of *reductio ad absurdum*, which Hegel distinguishes from the genuine dialectic (*internal with positive result*), prefigured by Plato. For Plato the contradictions enfolded within the analysis of the concept of "one" (the dialogue that, for Hegel, is the paradigmatic example of genuine dialectic is the *Parmenides*) do not destroy the concept, but are revelatory with respect to its complete determination and definition: the contradiction is the determination of what the concept is, the (formal) truth about it. In Aristotle's *Topics* Plato's dialectic is systematized and methodically articulated as logic of our thinking about *éndoxa* (the *éndoxa* are theses concerning controversial questions of universal interest such as: is justice the advantage of the stronger?). Hegel only hints at Aristotle's *Topics*, but he stresses, more generally, the fundamental continuity between Plato and Aristotle, and the genuinely speculative nature of Aristotle's philosophy. I share here the interpretation of those philosophers (in particular Berti) who see the continuity between Plato and Aristotle in the idea of dialectic as the *logic of philosophy* (whose aim is finding the truth about controversial theses pertaining to matters of universal interest). This meaning is evidently shared by Hegel himself.

In **Chapter 11** I consider the passage of the *Logic* in the *Encyclopaedia* (at the end of the "Preliminary Considerations") in which Hegel presents the three moments/sides of every *Logisch-Reelles* or of every *Logisches*. The passage is fundamental for two reasons: it contains Hegel's own definition of the formal structure of every dialectical and speculative inference, and presents the idea that this dialectical structure corresponds to the behaviour and method of every true thought, of truth. Hegel states that it is the clarification of the structure of everything that is "logical-real", and this means of conceptual thought, i.e. of the (second order) *thought about the forms of thought expressing the nature of reality*. In my view, this can be taken to be Hegel's own theory of dialectical validity.

In **Chapter 12** I reconsider Hegel's theory of dialectic focusing on its logical meaning (dialectic as the logic of contradiction), trying to answer the question "what is Hegel's dialectic?". I locate first of all Hegel's analyses in the *Lectures* in the context of the history of logic. I claim that Hegel's theory of dialectic belongs in all respects to the discussions on and the history of two kinds of inferences (which, in turn, are related to each other): *consequentia mirabilis* arguments and paradoxical deductions. In particular, I suggest that the passages of Hegel's *Lectures* on Megarian and Platonic dialectic give essential insights to highlight the closeness, but also the differences, between dialectic and these two kinds of inferences. In the chapter on the Megarians Hegel claims that the Liar paradox (in the interrogative form: "does the person who says that she lies lie or tell the truth?) cannot be solved univocally, answering the question only stating "she lies" or "she tells the truth". The paradox demands, according to Hegel, a double answer, that is: "she lies if she tells the truth and tells the truth if she lies" or also "she lies and tells the truth". For Hegel, this conjunction of contradictories is the truth, the true answer, while the single sentences "she lies", "she tells the truth" are untrue, or only partially true. In the passage on Plato Hegel highlights that, when we consider the concepts "per se" (i.e. we ask about the meaning of the concepts/predicates that we use in our sentences – for example the predicate "being"), "something marvellous [*das Wunderbare*] happens", namely the conceptual determination turns into its opposite ("being" turns into "non-being"). The same "marvellous" event happens when we consider the meaning of the determination/predicate "non-being": it turns into its opposite. Hence Hegel also stresses that the concept is actually the *movement* between the two conceptual determinations, while the determinations: "being", "non-being", singularly taken, are not sufficient to express the totality of the concept. On the basis of this analysis, I argue that Hegel's *Wunderbares*, terminologically and content-wise, belongs to the history of *consequentia mirabilis*. In both cases we derive α from $\neg\alpha$ or vice versa, but while in a *consequentia mirabilis*-argument the derivation of α from $\neg\alpha$ means that α is necessarily true, in dialectical inferences what is true is the passage from α to $\neg\alpha$ and vice versa. In this sense dialectical inferences can be likened to paradoxical deductions (in the meaning I derive from Routley and Priest of: a valid argument conveying a *true contradiction*). I consider the difference between the meaning of contradiction in Hegel and Priest and Routley in Part V. Here I just hint at it: I suggest that, in Hegel, the contradiction has a specific nature, it is to be formally expressed as $\alpha \leftrightarrow \neg\alpha$ (and not as $\alpha \wedge \neg\alpha$), whereby the biconditional cannot be separated ($\alpha \leftrightarrow \neg\alpha$ is true, but $\alpha \to \neg\alpha$ and $\neg\alpha \to \alpha$ cannot be stated separately).

At the end of Chapter 12 I draw some conclusions about Hegel's notion of validity compared to the contemporary one. I summarise some acquisitions and distinctions concerning the concept of validity or logical consequence: the distinction between semantic and syntactic, proof-theoretic and model-theoretic validity. The contemporary approach that best grasps Hegel's account is possibly the one of *semantic validity*. Hegel's theories do not intend to question the concept of validity as *truth preservation in virtue of form*. They rather intend to root it in the actual truth of our premises and conclusions, actual truth that, for Hegel, is conceptual truth.

V Contradiction

The unification [*die Vereinigung*] through which the elements of an antinomy are joined, is the norm. (Hegel Werke 1, 250–251)

The notion of contradiction at the core of dialectic is a controversial point in the reception of Hegel's thought.[1] Classically, interpreters either consider Hegel's philosophy as a serious enterprise and thus deny that Hegel's critique of the Law of Non-Contradiction (LNC) should be taken seriously, or they take this critique as a serious argument, and therefore deny that Hegel's philosophy should be taken seriously.

According to a widespread view, whose most influential exponent is probably Karl Popper, Hegel's dialectic is unscientific because it implies an "attack upon the Law of Non-Contradiction":

> [Hegel's idea of the fertility of contradictions] amounts to an attack upon the 'law of contradiction' [...] of traditional logic, a law which asserts that two contradictory statements can never be true together, or that a statement consisting of the conjunction of two contradictory statements must always be rejected as false on purely logical grounds [For this reason] if we are prepared [like Hegel] to put up with contradictions, criticism, and with it all intellectual progress, must come to an end. (Popper 1965, 16–17)

On a similar note, Charles S. Peirce observes: "As far as I know, Hegelians profess to be self-contradictory" (Peirce 1868, 57).

On the other side, many commentators deny that Hegel criticised LNC. In so doing, they try to save dialectic from the charge of being irrational and unscientific. According to McTaggart,

---

[1] The controversy starts with the debates on the so called "logische Frage". The expression "logical question" (*die logische Frage*) is commonly used to refer to the critical discussions about the Hegelian account of logic started in the first half of the 19th century. See on this Peckhaus 1997, Id. 2004, 3–14, Id. 2013, 283–296 and Lejeune 2013. Peckhaus 2013, 283, reconstructs that Trendelenburg's critique of Hegel was at the origin of the debate about the foundations of logic, a debate that had important implications for both the birth of Frege's logic and a new determination of the tasks and methods of philosophy itself. For a reconstruction of the discussions about the meaning of contradiction in Hegel's dialectics immediately after Hegel's death see also Colombo 1998, Spaventa 1972, 367–437, Merker 1951, Berti 1977, 9–31 and 161–181, Verra 1976, 13–38, Burkhard 1993, Wagner 2011, 23 ff., Ficara 2015, 39–55. On Trendelenburg's claims against Hegel see in particular Spaventa 1972, 392–405 and 412–437. Wolff 1981, 170 writes that the aim of Hegel's logic is to analyse "what contradictions are genuine contradictions". His hermeneutical hypothesis is that the classical logical laws are kept within Hegel's logic, and that Hegel's doctrine of *Widerspruch* (programmatically) cancels the difference between contrariety and contradiction. In this sense, Wolff does not question Trendelenburg's view about the reduction, in Hegel's system, of contrariety to contradiction, but rather tries to explain its philosophical and logical reasons, as well as its historical roots. In the last 10 years many works on Hegel's notion of contradiction have been published, among them see Hahn 2007, De Boer 2010, 345–373, Schick 2010, Illetterati 2010, 85–114 and Id. 2014, 127–152, Bordignon 2014.

> If the dialectic rejected the LNC, it would reduce itself to an absurdity, by rendering all argument, and even all assertion, unmeaning [...]. In fact, so far is the dialectic from denying the LNC, that it is especially based on it. (McTaggart 2000, 15)

More recently, a new hermeneutical line has gained greater importance. Brandom claims that "Hegel radicalizes LNC and places it at the very center of his thought" (Brandom 2002, 179). Similarly, according to Pippin, Hanna and Stewart "that Hegel rejected LNC" is "a Myth" which has to be revised.[2]

Significantly, if one adopts Popper's attitude, it is impossible to give an account of Hegel's specific meaning of "dialectic" as: "the fundamental tool in order to distinguish truth from falsity" (Eckermann 1987, 622–623.). By contrast, if one claims that Hegel did not criticise LNC, since when he spoke of "contradictions" he intended something else, one can hardly give an account of Hegelian claims such as "the LNC has no formal value for reason" (Hegel Werke 2, 230) or *"contradictio est regula veri, non contradictio falsi"* (Hegel Werke 2, 533).

In this scenario the paraconsistent approach, or the perspective according to which the admission of a contradiction does not imply the "explosion" of logic and rationality, plays a fundamental role.[3] My hypothesis is that Hegel does not

---

[2] See the chapter on the "Myth that Hegel Rejected the Law of Non-Contradiction" in Stewart (ed.) 1996, 38–84.
[3] For a paraconsistent reading of dialectic see Marconi 1979b, 46 ff., Apostel 1979, 85–113, Routley/Meyer 1979, 324–354. For a dialetheist reading see Priest 1989, 388–415, and more recently Ficara 2013, 35–52 and Bordignon 2014. For Günther 1978, Vff. in Hegel emerges a new form of rationality, which breaks with classical bivalent logic and Günther calls "trans-classical". Günther 1978, VIII and IX stresses that the attempts, started with Łukasievicz 1920, and anticipated by Peirce in 1909, of establishing a trivalent or plurivalent logic do not really manage to overcome the classical dualistic frame. For Günther these approaches do not question the polarity of "true" and "false" insofar as they posit the "third" values as a mere *"between"* between true and false. In Hegel, in contrast, we have the idea of a genuine mediation (*Vermittlung*) of the two poles. Günther himself admits that his attempts at explaining the logical meaning of "*Vermittlung*", in some way, failed. The core of the problem is, as Günther 1978, X claims, the meaning of negation and double negation in Hegel. On the difficulty of Günther's account of dialectics see Marconi 1979b, 29 f. However, it is difficult to admit that we have achieved now an adequate formal account of dialectical inferences, and the discussions are still open. Günther 1978, XV also claims that his main concern in presenting Hegel's view is not primarily exegetical but meta-scientific. In other words, he wants to show how contemporary conceptions of rationality completely ignore an enormous field of research, the field of "transclassical rationality" discovered not only by Hegel, but by the whole idealistic tradition. Toth 1987, 89–182 analyses the history of geometry and interprets the relation between Euclidean and non-Euclidean geometry in dialectical terms. In his reconstruction, the two sciences are both true, and one is the negation of the other. Their emergence in history postulates the existence of what Hegel called "reason", the horizon which mediates between the two and contains both. *Vernunft* coincides for Toth with a

reject *every* form of LNC, but only a specific form of it. This is in fundamental accordance with a recent tendency within paraconsistent approaches of specifically dialetheic inspiration, according to which some versions of LNC are acceptable.[4] Thus the contemporary treatment of contradictions helps us to shed light on Hegel's critique of LNC. However, on closer examination we will see that Hegel's view significantly differs from the dialetheic account on some interesting points.

In what follows I will give a closer look at the connectives (conjunction and negation) and correspondent logical laws (Double Negation Elimination, LNC, LEM, Simplification) questioned by or involved in dialectical contradictions. I first analyse what Hegel says on conjunction (13.), negation (14.) and contradiction (15.). In the final chapter (16.), I consider the relevance of Hegel's account for debates on contradictions in philosophical logic.

---

meta-metalinguistic realm. In this light, dialectical logic corresponds to a non-classical logic that admits of four truth-values: true, false, neither and both.

4 See Berto 2007a, Chapter 1.

13 Conjunction [*Vereinigung*]

While literature abounds on Hegel's concept of negation, there are no works monographically devoted to the connective that joins the two elements of a dialectical contradiction.[5] And yet, as I will show in what follows, Hegel formulates a specific theory about the link joining the two terms of an antinomy. In his early writings he calls this connective *Vereinigung*. This use and the underlying concept is confirmed in his later works. The German term *Vereinigung* means, literally, "unification". In Hegel's use *Vereinigung* has a logical meaning: it refers to the connective joining the two theses of an antinomy. Besides this, it also has other implications: epistemological (in designating the activity of unifying the opposites of the antinomy in one single belief) and theological (in that the belief Hegel writes about is religious belief, and its content is God as contradictory con-

---

[5] Hints at the semantics of conjunction in dialectical contradictions can be found in Wetter 1958, Havas 1981, Priest 1989, 388–415 and Bordignon 2014, 87. Wetter 1958, 340 observes that "the [dialectical] opposites are so far intertwined that the one cannot exist without the other". Havas 1981, 257–264 discusses the account of Hegel's dialectic given by Routley/Meyer 1976, 1–25 and in particular their claim according to which dialectical conjunction works classically. Havas 1981, 258 writes (I change Havas' notation to bring it in line with my usage): "The truth functional connective $\wedge$ is inadequate to represent the dialectical unity of opposites [...] the truth of $a \vee \neg a$ does not follow from $a \wedge \neg a$ if this latter proposition happens to be the representation of a dialectical contradiction. Dialectically contradictory aspects can exist, and can be true, only together". Similarly, Priest 1989, 396–397 suggests that dialectical true contradictions imply a stronger kind of conjunction between a sentence and its negation than a merely extensional one (i.e., standard truth-functional conjunction). Accordingly, Priest 1989, 396–397 claims that "there should be a more intimate relation between dialectical contradictories than the mere extensional (external) conjunction". Paraconsistent philosophers (e.g., Meyer, Priest, Routley) have, accordingly, advocated an "intensional" conjunction (which is common in relevance logics), and which fails to obey simplification or, in many cases, adjunction. Also Bordignon 2014, 87–88 highlights some characteristic aspects of Hegelian contradictions, writing that "from the expression of the truth of Becoming [...] it is not possible to deduce the truth of the single conjuncts [Being and Nothing]" and also that "in the linguistic expression of the contradictory nature of Becoming the truth of *p (Being and Nothing are the same)* does not imply the falsity of *not-p (Being and Nothing are not the same)*: *p* and *not-p* are both true and true in their unity". I agree with the spirit of such previous work: Hegelian contradictories are linked in a way that is stronger than common conjunction. Yet my account is different, for two main reasons. First, I defend the insight that while the dialectical contradiction is true, the two contradictory elements joined in a dialectical contradiction are not true (for the focus on this insight in dealing with the logic of dialectical contradictions I am grateful to Franca d'Agostini and Jc Beall). Second, I stress that dialectical contradictions are not *conjunctive* but rather *biconditional* contradictions, they have the form *a iff not-a*, from which we cannot logically derive the conjunctive contradiction *a and not-a*.

cept). In what follows, I will focus on the logical meaning, translating it as "conjunction" rather than "unification". I will also show that the link Hegel is writing about when he writes about *Vereinigung* is stronger than the one established by a simple logical conjunction. Logically, its behaviour is rather expressible in terms of a *biconditional* whose antecedent is a sentence ("matter is continuous") expressing a conceptual determination ("matter's continuity"), and the consequent its negation ("matter is not continuous"), expressing the opposite conceptual determination ("matter's punctuality"). In this light, Hegel's theory of *Vereinigung* perfectly explains the meaning of dialectical contradictions, allowing a genuine assessment of their place within the history of paraconsistent logic.

In the fragment on *Glauben und Sein*, written in Frankfurt between December 1797 and the beginning of 1798 Hegel writes that *Vereinigung* is:

> [T]he conjunction [*Vereinigung* – unification] through which an antinomy is conjoined [...] In order to be conjoined (*vereinigt* – unified) the elements of an antinomy have to be recognised as contradictorily opposed to each other, their relation between each other has to be felt and known as an antinomy; but the opposite can be recognised as an opposite only if it has already been unified (*vereinigt*); the conjunction (*Vereinigung*) is the norm through which the comparison takes place and through which the opposites, as such, emerge in their insufficiency. (Hegel Werke 1, 251)

Many authors suggest that the fragment represents the germ cell of Hegel's dialectical method, which is systematically developed in Hegel's mature writings.[6] We find here the focus on the logical importance of conjunction (*Vereinigung*) in Hegel's logic.

That "conjunction is the norm" means that the truth about the content of a conceptual given can be found only in the conjunction of the opposites and that the elements of an antinomy, taken in isolation, are "insufficient", i.e. not able to express *on their own* the whole content at stake. In order to express the whole content they need to be joined with their contradictory opposites.

In the *Differenzschrift* Hegel also talks about the "principle of completion" (*das Prinzip der Vervollständigung*), which consists in:

> [C]ompleting the limitations [i.e. the single thesis or the single antithesis of an antinomy] by affirming their contradictories, as their conditions; the latter limitations need, in turn, the same completion. (Hegel Werke 2, 26)

---

[6] However, there are also specific differences between Hegel's early theory of *Vereinigung* and his later dialectical view. See on this Pöggeler 1981, 42–45, Düsing 2012, 11–114 among others.

Hegel here uses the word *Vereinigung* in order to denote a specific conjunction, namely the one of the thesis and the antithesis in an antinomy. *Vereinigung* is then the conjunction within the contradiction (p and not-p), and not a conjunction of two merely different propositions (p and q). What is more, a single thesis in an antinomy ("matter is continuous"), as Hegel writes, *needs to be completed by its contradictory as its condition*. Clearly, the link joining the two contradictory theses is not identical to a mere conjunction. The two theses in an antinomy are not simply stated paratactically, but they necessitate each other.

This explains the basic idea of what one can call *Hegelian contradiction:*

In a true contradiction, the contradictory terms cannot be separated, as they, in isolation, are insufficient to give a full account of the content at stake.

In the *Lectures on the History of Philosophy* Hegel, recalling Plato, distinguishes between dialectic and sophistic, writing that the sophists produce contradictions, but "*do not bring these thoughts together*" and "*do not really unite the opposites*", but rather "separate all existences from one another [causing] everything to fall asunder" (Hegel Werke 19, 76/Hegel 1892 ff., vol. II., 68). In contrast dialectic consists in "showing that what is the other is the same, and what is the same, is another, and likewise in the same regard and from the same point of view" (Hegel Werke 19, 76/Hegel 1892 ff., vol. II., 68).

Hence Hegel distinguishes between two kinds of *unities of the opposites:* the sophistic and the dialectical one. The opposites (bitter and sweet, large and small etc.) are not truly put together by the sophists. In fact the opposites the sophists refer to ("honey is sweet and honey is not sweet" "six is great and six is small") can be truly affirmed separately (six is great against four but small against eight). In contrast, a dialectician holds that, if p is a true contradiction, you cannot affirm p without not-p and vice versa. Affirming only one of the two conjuncts would be wrong.

In the *Lectures on the History of Philosophy* Hegel, as we have seen, considers Megarian philosophy, and in particular Eubulide's paradoxes, as examples of dialectical developments. Interestingly, here too conjunction/*Vereinigung* plays a fundamental role. Discussing the Liar paradox (in its interrogative form: "if someone says s/he lies, does s/he lie or tell the truth?"), Hegel writes that "a simple answer is demanded", but "cannot be given" because "here we have a union of two opposites, lying and truth, and their immediate contradiction". Those who require a simple answer "seek a simple relation from something incommensurable, i.e. they fall into the error of demanding a simple reply where the content is contradictory" (Hegel Werke 18, 529/Hegel 1892 ff., vol. I., 459 f.).

If we focus on the behaviour of the conjunction, Hegel's discussion implies that while the simple proposition p (the Liar tells the truth) is the wrong answer and the other simple proposition: not-p (the Liar lies) is wrong, the compound sentence: "p and not-p" is true ("he thus both lies and does not lie"). This happens because the content at stake, according to Hegel, is "the union of two opposites", "incommensurable" and "contradictory" while the sentences aiming to describe it are "simple". The same holds for the following two cases, mentioned by Hegel in the *Introduction* to the *Lectures on the History of Philosophy*:

> We do not controvert the fact, or think it contradictory, that the smell and taste of the flower, although otherwise opposed, are yet clearly in one subject; nor do we place the one against the other. But the understanding and understanding thought find everything of a different kind, placed together, to be incompatible. Matter, for example, is complex [...] or space is continuous and uninterrupted. Likewise we may take separate points in space and break up matter dividing it ever further into infinity. It then is said that matter consists of atoms and points, and hence is not continuous. Therefore we have here the two determinations of continuity and of definite points, which understanding regards as mutually exclusive, combined in one. It is said that matter must be clearly either continuous or divisible into points, but in reality it has both these qualities. (Hegel Werke 18, 44/ Hegel 1892ff., vol. I., 26)

And:

> [W]hen we say of the human mind that it has freedom, the understanding at once brings up the other quality, which in this case is necessity, saying that if mind is free it is not in subjection to necessity, and, inversely, if its will and thought are determined through necessity, it is not free – the one, they say, excludes the other. The distinctions here are regarded as exclusive, and not as forming something concrete. But that which is true, the mind, is concrete, and its attributes are freedom and necessity. Similarly the higher point of view is that mind is free in its necessity, and finds its freedom in it alone, since its necessity rests on its freedom (Hegel Werke 18, 45/Hegel 1892ff., vol. I., 26).

While the sentences "matter is continuous" and "matter is not continuous", "the human mind is free" and "the human mind is not free" are, if taken individually, untrue, their connection: "matter is continuous and not continuous"; "the human mind is free in its necessity, and subjected to necessity in its freedom" are true. We also see that Hegel refers here to one concept (the concept of *Geist*/mind, the concept of matter) and sees the conjunction of the incompatible determinations as the complete determination of its content.

In the second note to the chapter about *Being* in the *Science of Logic*, Hegel addresses the question of the link between truth and contradiction by focusing on the role of *Vereinigung*. He first stresses that the result of the considering the concept of being in the first chapter is the insight that: "being and nothing are the same".

> Now in so far as the proposition: "being and nothing are the same", asserts the identity of these determinations, but, in fact, equally contains them both as distinguished, the proposition is self-contradictory and cancels itself out. Bearing this in mind and looking at the proposition more closely, we find that it has a movement which involves the spontaneous vanishing of the proposition itself. But in thus vanishing, there takes place in it that which is to constitute its own peculiar content, namely, becoming. (Hegel Werke 5, 92–93/Hegel 1969, 90)

Hegel says that, if we look more closely at the structure and the content of the sentence, we realise that it contradicts itself. What it "contains" are two opposite determinations, being and nothing, and what it states is that they are not opposite at all. This means that the sentence "has a movement", i.e. the spontaneous vanishing of the proposition itself. In other words, insofar as we state it we are doomed to deny it. But this happens because we look at it from a closer standpoint, and reflect about it:

> The sentence thus *entails* the result [...] but the fact to which we must pay attention here is the defect that the result is not itself *expressed* in the sentence; it is an external reflection which discerns it therein. (Hegel Werke 5, 92–93/Hegel 1969, 90)

Hegel also stresses that this (the sentence implicitly entails its negation, but does not express it) is a defect, which depends on the fact that we try to express what is complex (speculative) using a single statement:

> In this connection we must, at the outset, make this general observation, namely, that the sentence *in the form of a judgment* (*Urteil*) is not suited to express speculative truths; a familiarity with this fact is likely to remove many misunderstandings of speculative truths. Judgment is an *identical* relation between subject and predicate; in it we abstract from the fact that the subject has a number of determinations other than that of the predicate, and also that the predicate is more extensive than the subject. Now if the content is speculative, then the *non-identical aspect* of subject and predicate is also an essential moment, but in the judgment this is not expressed. (Hegel Werke 5, 93/Hegel 1969, 90f.)

Hegel hints here at the fact that if we interpret the copula in a sentence as an expression of identity, which we have to do when we formulate judgements, i.e. definitional sentences (see on this Part III), stating for instance "justice is the advantage of the stronger", we have to abstract from the fact that justice also has other properties than the one expressed by the predicate "being the advantage of the stronger". If the content is speculative, the non-identity, or logical negation of the original statement, plays a fundamental role.

If the content is speculative, then the simple single sentence is insufficient to express the whole content. Speculative means: entailing its (contradictory) opposite, its negation. The sentence expresses only one side of the speculative content, namely the identity, and leaves out the non-identity between being and nothing. But, Hegel says, there is a way to correct or complete the insufficiency, and to express the speculative truth.

> To help express the speculative truth, the deficiency is made good in the first place by adding the opposite sentence: "being and nothing are not the same", which is also enunciated as above. (Hegel Werke 5, 94/Hegel 1969, 458)

Evidently, Hegel is talking here about two *contradictorily* opposite sentences. In what follows he explicitly refers to the fact that "completing" the first sentence with its opposite generates an antinomy. As above, Hegel suggests that, when the content is speculative (and this means heterogeneous, complex, incommensurable) one single sentence is insufficient to express it, and needs to be completed through its negation. But then a further problem emerges:

> But thus there arises the further defect that these propositions are not connected (*verbunden*), and therefore exhibit their content only in the form of an antinomy, whereas their content rather refers to one and the same thing, and the determinations which are expressed in the two propositions are supposed to be plainly conjoined (*vereinigt*) – a conjunction (*Vereinigung*) which can only be expressed as an *unrest* of *incompatibles*, as a *movement*. (Hegel Werke 5, 94/Hegel 1969, 458)

Hence, whereas the two theses of an antinomy are disconnected, the two opposite statements expressing a speculative content need to be connected. More specifically, they need to be connected through a conjunction that is also *unrest* and *movement*. What the "movement" is emerges in the following passage:

> The commonest injustice done to a speculative content is to make it one-sided, that is, to give prominence only to one of the propositions into which it can be resolved. It cannot then be denied that this proposition is asserted; *but the statement is just as false as it is true*, for once one of the propositions is taken out of the speculative content, the other must at least be equally considered and stated. (Hegel Werke 5, 94/Hegel 1969, 458)

Thus the "movement" refers to the fact that, once we assert the truth of one side of the speculative content, we cannot rest at it, but must assert the other (the negation of the first side) and vice versa. Again, the paradoxical structure of dialectical inferences comes into view: that the statement p ("being and nothing are the same") is just as false as true means that, since it entails its negation, ¬p ("being and nothing are not the same"), and, in turn, ¬a implies a, a is false if true and true if false. From this point of view, a more suitable expression of Hegel's *Vereinigung* would be the biconditional $a \leftrightarrow \neg a$.

14 Negation

Literature on negation in philosophical logic, as Wansing (2007, 415) has suggested, abounds with disagreement. Literature on what we could call *Hegelian negation* is perhaps less vast but similarly controversial.[7] A problem highlighted by many authors (*in primis* Henrich 1978) concerns the disproportion between the crucial role of *negativity* within dialectic[8] and the scarcity of statements on the meaning of *negation* in Hegel's works. This is possibly one reason why many authors stress the radical difference between the dialectical and the standard, logical approach to negation, warning that Hegelian negation should not be flattened on standard, classical negation (the operator which, if applied to a sentence, forms its contradictory, producing a new sentence that is true iff the first is false). Interestingly, Henrich complains about the fact that Hegel, presenting his own view on (double) negation, always uses the traditional terminology, and does not distinguish the dialectical from the traditional account (of *duplex negatio affirmat*). "Hegel" Henrich (1978, 224) writes "everywhere talks as if they [the classical and the Hegelian double negation] were one and the same form of negation" and does not distinguish between the two. In what follows I will try to show that Hegel *does* distinguish between his account of negation and other traditional accounts.[9]

It should be specified in advance that Hegelian negation is to be intended in both conceptual and propositional terms.[10] As we have seen, Hegel's logic is a conceptual logic, dealing with the task of analysing and determining the true meaning of concepts, but this does not mean that it is not also propositional or sentential. Clearly, Hegel's early theory of *Vereinigung* and its mature develop-

---

[7] Puntel 1996, 131–165 claims that Hegel's conception of dialectic, and more specifically *negation*, is "not intelligible". For Puntel dialectical negation leads to an infinite regress, and Hegel's claim that the dialectical method has a positive result is unsustainable. His conclusion is that Hegel's dialectic cannot pretend to be an acceptable explication of the "intuitive" understanding of negation.

[8] The famous *ungeheure Macht des Negativen* (Hegel Werke 3, 36/Hegel 1977, 19).

[9] More generally, there seems to be a certain incommunicability between the treatment of negation in the history of logic and its consideration in the history of philosophy, partially reflected by the presence of two articles on "negation" in the *Historisches Wörterbuch der Philosophie*, the first on negation from a logical point of view, the second on negation and negativity in philosophy. See Ritter/Gründer/Gabriel (eds.) 1971ff., vol. 6, 666–671.

[10] On the difference between propositional and predicative uses of negation see Horn 1989 and Horn/Wansing 2016. Berto explains the legitimacy of the passage from the conceptual to the propositional and strictly logical dimension implied in Hegel's dialectics in 2005, 262ff. see also Berto 2007b, 19–39.

ment in the *Science of Logic* implies that the differences and oppositions within concepts (the concept of being, the concept of matter, the concept of human freedom) are to be expressed in terms of sentences.[11] Thus, when Hegel speaks of "negation" he intends both the negation of the predicate-concept (¬P), and the negation of the sentence that describes the nature of this predicate-concept (¬p). This means that when we speak of negation in Hegel's logic we are referring to the operator commonly used by logicians.

Moreover, Hegelian negation is to be classically intended as contradictory forming operator. This means that the two terms P and ¬P or p and ¬p are taken to be mutually exclusive and jointly exhaustive. This will come more clearly into view when I will consider the notion of contradiction.

Despite this, Hegelian negation differs from the usual (classical and non-classical) one in some interesting respects. Three points should be stressed that mark the specific nature of Hegelian negation with respect to both the classical and the non-classical logical tradition. The first is the idea that negation "falls within the content", i.e. is internal to conceptual content. The second is the idea that it has positive content, insofar as it is "the positive substance of that content". The special, determining character of Hegelian negation introduces the third aspect, Hegel's new conception of double negation, strictly related to contradiction: according to it, applying negation to itself does not produce affirmation, but contradiction. These aspects may seem counter-intuitive. In particular, it seems difficult to understand how negation for Hegel can be internal and positive, as well as expression of a contradictory relation between sentences or predicates. However, in light of Hegel's theory of *Vereinigung*, and of my interpretation of dialectical unities of opposites as biconditionals of the form $a \leftrightarrow \neg a$, all the different aspects mentioned in Hegel's semantics of negation are perfectly understandable.

These aspects are also expressed by Hegel in terms of the famous concept of *determinate negation*.[12] In the *Preface* to the *Phenomenology of Spirit* Hegel writes:

---

[11] In today's terminology, it is common to refer to the sentences that make the meaning of dialectical conceptual contents explicit as meaning postulates. See Berto 2005, Chapter VIII. According to Berto, Hegel's dialectics deals with relations between conceptual determinations, and these relations are expressed by "implicative sentences" or "meaning postulates".

[12] Hegel's notion of determinate negation is the topic of classical works. Among them are the already mentioned Henrich 1978, as well as Düsing 2012 (Chapter 1), Cortella 1995, Landucci 1978, Perelda 2003, Redding 2007 (Chapter 3), Viellard-Baron 2013, 46–68, Pippin 2014, 87–110. Among the works on the link between Hegel's negation and the non-classical logical tradition the essays collected in Marconi (ed.) 1979a are worth mentioning. Berto 2005, 284 ff. exam-

> To see what the content is *not* is merely a negative process [...] it is the negative with no awareness about the positive element within it [...] On the other hand, in the case of conceptual thinking [...] the negative aspect falls within the content itself, and is the positive substance of that content [...] Looked at as a result, it is the *determinate* negative, the negative which is the outcome of this process, and consequently is a positive content as well. (Hegel Werke 3, 57/Hegel 1977, 36)

The negation of a conceptual determination (such as the negation of "matter is continuous") "falls within" the concept of matter, which contains both continuity and its negation. From the point of view of my interpretation of dialectical true contradictions in terms of $a \leftrightarrow \neg a$, that negation is internal means that it is within the double implication (which is the full and adequate expression of the concept), and not outside of it. Let us consider the example mentioned by Hegel in the *Introduction* to the *Lectures on the History of Philosophy*. If we grant that the concept of human mind is fully and adequately expressed by the biconditional "mind is free in its necessity, and finds its freedom in it alone, since its necessity rests on its freedom" (mind is free if subjected to necessity and subjected to necessity if free, whereby "subjected to necessity" is to be taken as the negation of "being free"), then the negative "the human mind is not free" is simply a part of the fully developed concept, a determination within it.

As to the second point, Hegel assumes that $\neg P$ and $\neg p$ have (in speculative logic) a *positive result:* they are not elimination or absence of P nor non-subsistence of p. In the introduction to the *Phenomenology of Spirit* Hegel, distinguishing his view on negation from Pyrrhonian negation, observes:

---

ines the relationship between holistic inferentialism and Hegel's dialectics, focusing in particular on the meaning of dialectical negation as material incompatibility. In Ficara 2014a, 29–38 I highlight the similarities between Hegel's determinate negation and glutty negation. Wolff 1986, 107–128 highlights that the background of Hegel's notion of negativity is the Kantian distinction between logical and real opposition. On Hegel's and Kant's notion of negation see also more extensively Wolff 1981. Bordignon (2014, 35) writes that applying the Kantian distinction between logical and real opposition to Hegel's logic is misleading because "it does not take into account that Hegel's logical system is informed by the identity of being and thought". Yet, Bordignon (2014, 61) also states that Hegel's determinate negation "does not pertain to the linguistic-propositional realm, but to the ontological one. Consequently, determinate negation has nothing to do with the truth of propositions or with the subsistence of determinate states of affairs, but rather with the dynamics at the basis of the same articulation of reality". I highlight, in contrast, that dialectical negation is the logical and linguistic expression of ontological negativity, i.e. of incompatibility relations between conceptual determinations, relations that are expressible through sentences and involve a standard meaning of negation as contradictory forming operator.

> This is just the scepticism, which only ever sees the *pure nothing* in its result, and abstracts from the fact that this nothing is determinate, the nothing of *that from which it results*. Nothing, however, is only, in fact, the true result, when taken as the nothing of what it comes from; it is thus itself a determinate nothing, and has a *content*. The scepticism which ends with the abstraction "nothing" or "emptiness" can advance from this not a step farther, but must wait and see whether there is possibly anything new offered, and what that is – in order to cast it into the same abysmal void. When once, on the other hand, the result is apprehended, as it truly is, as *determinate* negation, a new form has thereby immediately arisen; and in the negation the transition is made by which the progress through the complete succession of forms comes about of itself. (Hegel Werke 3, 74/ Hegel 1977, 51)

And:

> We have here, however, the same sort of circumstance, again, of which we spoke a short time ago when dealing with the relation of this exposition to scepticism, viz. that the result which at any time comes about in the case of an untrue mode of knowledge cannot possibly collapse into an empty nothing, but must necessarily be taken as the nothing *of that of which it is a result* – a result which contains what truth the preceding mode of knowledge has in it. (Hegel Werke 3, 79–80/Hegel 1977, 56)

By "scepticism" Hegel here means, evidently, the ancient (Pyrrhonian) insight that we have already considered, according to which for every valid argument there is an opposite one that is equally valid, or for every apparently true sentence there is an opposite one that is equally true (*panti logo logos isos antikeitai*). Hegel writes about "the *negation* of that of which it is a result...", suggesting that the form of the sceptical principle is simply a contradiction:

$$p, \neg p$$

The sceptic "sees in the result only pure nothingness" and "the result collapses into an empty nothing". In other words: the sceptic holds that a contradiction entails nothing, the disappearance of things, and any knowledge of them. This means that the sceptical view of negation is cancellation: stating $\neg p$ means nullifying or cancelling p.[13] The dialectician instead admits the fact "that this nothing is determinate and has a content", and sees "the result as it truly is, as *determinate* negation" and as "containing what truth the preceding mode of knowledge has in it".

---

13 On the meaning of negation as cancellation see Priest 2006, 75 ff. On negation as cancellation in connexivist logics see Routley 1978, 393–412.

That negation is positive can be seen by recalling the description of dialectical inferences in the previous part and from the point of view of my interpretation of Hegelian contradictions as biconditionals of the form $a \leftrightarrow \neg a$. The dialectical task of determining the meaning of concepts such as "matter" or "the one" involves that every determination concerning them forces us to infer its negation. From "matter is continuous" we infer "matter is not continuous", from "the one is one" we infer "the one is not one". The negatives: "matter is not continuous", "the one is not one" are said to be positive, or as much positive as the corresponding affirmations because from "matter is not continuous" we are forced to logically infer "matter is continuous". Hence the negation of p does not cancel p out, but it simply reproduces it.

Finally, the consequence of speculative negation is contradiction. Hegel says this in claiming that what follows from the negation is not nothing but rather the truth of the preceding mode of knowledge. The "truth of the preceding mode of knowledge" is simply the recognition of the truth of the contradiction at stake. There have been different interpretations of the dialectical process in virtue of which the various moments are overcome and maintained. Logically speaking, this sort of positive and maintaining negation can be interpreted in terms of double negation. Hegel himself supports this interpretation, when he explicitly states, as we will see, that "negation of negation is contradiction" (Hegel Werke 20, 164/Hegel 1892ff., vol. III., 256ff.).[14]

More specifically, it is the special (determining, positive) nature of Hegelian negation that also introduces a new conception of double negation strictly related to contradiction. In the introduction to the *Science of Logic* Hegel condenses his ideas about negation:

> In the *Phenomenology of Spirit* I have expounded an example of this method in application to a more concrete object, namely to consciousness [...] *All that is necessary to achieve scientific progress* – and it is essential to strive to gain this quite *simple* insight – is the recognition of the logical principle that the negative is just as much positive, or that what is con-

---

[14] The statement "negation of negation is contradiction" is not contained in the English translation. For an interpretation of dialectical double negation in terms of generation of a contradiction see Düsing 2012, 50 and Baum 1986, 65–76. According to Düsing 2012, 50 "In one argumentative process one conceptual determination is affirmed, one contrary property is opposed to the first, this opposite determination is further turned into the contradictory opposite of the first [...] the negation of the negation turns to the whole [concept] that [...] entails that opposition and contradiction". "So wird in einem und demselben Argumentationsgang eine Bestimmung gesetzt, eine ihr konträr entgegengesetzte ihr gegenübergesetzt, dieser inhaltlich bestimmte Gegensatz zum Widerspruch [...] fortbestimmt [...] Die Negation dieser Negation führt positiv zurück auf das zugrunde liegende Ganze, das [...] jenen Gegensatz und Widerspruch in sich bewahrt".

tradictory does not resolve itself into a nullity, into abstract nothingness, but essentially only into the negation of its *particular* content, in other words, that such a negation is not all negation but the *negation of a determinate subject matter* [bestimmte Sache] which vanishes, and consequently is a determinate negation [bestimmte Negation] and therefore the result contains that from which it results. (Hegel Werke 5, 49/Hegel 1969, 54)

The only "logical principle" accepted by Hegel (who, as we will see, generally rejects *Grundsatzphilosophien* – every attempt at fixing one principle as the basis of philosophical knowledge) is the following: "the negative is just as much positive" which is the same as "what is contradictory does not resolve itself into nullity". By "negative" Hegel thus means here "contradictory", and by determinate negation the specific attitude towards contradictions implied by dialectic.

Here we therefore see that the contradictory sentences "not-p" and "p" "have particular contents", by which we ought to understand that they are *partial truths* about the concept P and the conceptual fact p that the sentence "p" is intended to capture. In other words, we have a concept, P, and we have to describe it. In describing it, we find its negation, ¬P, which is not nothing, but as much as positive as the positive P; at this point we have both P and ¬P and our description of the conceptual fact at stake will be the contradiction.

The conception of determinate negation and the meaning of Hegelian double negation is further clarified in the chapter on Spinoza in the *Lectures on the History of Philosophy*, were Hegel distinguishes dialectical negation from Spinoza's principle *determinatio est negatio*.[15] Here, it also clearly emerges that approaches identifying Hegelian with Spinozian negation are misleading:

> Spinoza's procedure is therefore quite correct; yet [his principle] is false, seeing that it expresses only one side of the negation. The understanding has determinations which do not contradict one another; contradiction the understanding cannot suffer. The negation of negation is, however, contradiction, for in that it negates negation as simple determination, it is on the one hand affirmation, but on the other hand also really negation; and this contradiction, which is a matter pertaining to reason, is lacking in the case of Spinoza. (Hegel Werke 20, 164/Hegel 1892 ff., vol. III., 262)

In a word, Spinoza's double negation is intellectual and not rational: so it cannot suffer any contradiction. Spinoza's formula *determinatio est negatio* according to Hegel expresses only "one side of the negation". In Spinoza's view while the sub-

---

15 Spinoza's original formulation: "...et determinatio negatio est" is changed by Hegel into "omnis determinatio est negatio". See Jacobi 1998 ff., vol. 1.1, 100 and Hegel's review: Hegel Werke 4, 429–461. On the role of Spinoza in Hegel and Jacobi see Sandkaulen 2019, Chapter 14.

stance is pure affirmation, its determinations are negations (because they introduce cuts and limits within the infinite and continuous substance), they have a positive meaning insofar as they determine the substance *ex negativo*. However, the definitions of the substance are negative, but exclude contradiction, they "don't contradict one another" and we cannot use contradictory determinations in order to express the infinite substance. But using contradictory determinations in order to find the true meaning of concepts is precisely what Hegel wants to do.

Hegel therefore points out that it is right to say that the substance is affirmative insofar as it is the negation of the single determinations, which are themselves negations. So the substance is affirmation in the classical strong sense of "negation of negation":

$a = \neg\neg\, a$

However, the fact that the substance is affirmation in the sense of "negation of negation" does not mean that it is *only* affirmation. On the contrary, according to Hegel, it is also negation.

Spinoza claims that the infinite and the substance are "absolute affirmations". According to Hegel, this is right insofar as affirmation is nothing other than negation of negation. However, according to Hegel the expression "negation of negation" is better than "affirmation". Why? Because what Hegel is talking about here is a concept, such as the concept of *causa sui*, or Spinoza's substance, whose definition, whose same meaning, is the contradiction $a \leftrightarrow \neg a$. Therefore, we can formulate the Law of Dialectical Double Negation (or Dialectical Determinate Negation=DDN), according to which:

DDN: $\neg\neg a \vdash a \leftrightarrow \neg a$

This special law is easily explained if one takes into account that the concept Hegel is speaking of when he says that determinate negation, differently from what Spinoza thought, "does not only affirm but also genuinely negates" is an internally contradictory concept. So if we call the concept "α"[16] then we have that: $= a \leftrightarrow \neg a$, and it follows that $\neg\neg a \vdash a \leftrightarrow \neg a$, dynamically expressed by Hegel by saying that we capture the true nature of concepts only when, by negating their negation, we gain them in their completeness, which is contradictory.

---

[16] Surely the difference is to be related to the fact that Spinoza is speaking about the completeness of being, whereas Hegel speaks about the completeness of concepts.

An example might make the reasons for DDN clear. Hegel often emphasises that Spinoza's concept of *causa sui* as that whose essence entails existence is a truly contradictory concept.[17] In the case of the concept of *causa sui*, the property of "being cause" entails that "a cause can only be cause of something else"; now we see that the possessive pronoun "sui" entails the negation of the term *causa*, thus *causa sui* is something which is both *causa* and *not-causa*, reflexive and not-reflexive, i.e. a contradiction. According to Hegel, this means that the concept of *causa sui* contains a negation and the negation of this negation. The two determinations (*causa* and *sui*) are incompatible and yet joined in a unique notion, which connects both in an inextricable way.

DDN is rooted in the interpretation of Hegelian *Vereinigung* as a biconditional with the form $a \leftrightarrow \neg a$. Negating $\neg a$ (stating for example "it is not the case that the human mind is not free") is indeed a more appropriate expression of the truth of the biconditional $a \leftrightarrow \neg a$ than simply affirming, or simply negating a. If the biconditional $a \leftrightarrow \neg a$ is an expression of the complete truth about the concept of the human mind, then $\neg a$ is not sufficient to express the whole truth, and $\neg\neg a$ has the right to be stated. However, from $\neg\neg a$ it does not follow that a, but that the contradiction $a \leftrightarrow \neg a$ is true.

---

17 See Hegel Werke 20, 171/Hegel 1892 ff., vol. III., 262–263 and Hegel Werke 2, 229–230.

15 The Law of Non-Contradiction and the Law of Excluded Middle

Hegel's theory of negation and *Vereinigung* already presents all the elements of his conception of contradiction. Now, I consider what Hegel says on contradiction with special reference to the laws that, in traditional Aristotelian logic, rule the relation between truth and contradiction, namely: the Principle of Identity (I), the Law of Non-Contradiction (LNC) and the Law of Excluded Middle (LEM).[18]

Hegel's treatment of *Vereinigung* is the germ cell of dialectic and anticipates, in many ways, his conception of dialectic's logical core, the notion of contradiction.[19] Hegel's view on the link between truth and contradiction is already fully developed in his first published writings (the writings of the Jena period), first of all the so-called *Differenzschrift* (1801) and the *Essay on Scepticism* (1802). Its mature exposition is to be found in the *Wesenslogik* ("Logic of Essence") in the *Science of Logic* and in the *Encyclopaedia Logic*. I will pay special attention to Hegel's early writings: the *Differenzschrift* and the *Skeptizismusaufsatz*. In them the idea (fundamental for my purposes in this book) of *contradiction as the form of truth* is presented in the clearest terms, an idea that anticipates Hegel's treatment

---

18 On Hegel's view on LNC and LEM see Stekeler-Weithofer 1992, 23 ff. For Stekeler-Weithofer Hegel does not question the validity of LNC and LEM, but rather criticises "the *formalistic* assumption according to which these principles hold in general". As Stekeler-Weithofer writes, the two principles hold for declarative sentences, i.e. meaningful linguistic expressions in which we affirm or deny something of something, and which can be true or false. However, LNC and LEM do not hold for special sentences, i.e. liar-like sentences, or also preliminary formulations of a problem and articulations of meaning. More specifically, they do not hold for speculative sentences, i.e., for Stekeler-Weithofer 1992, 24, "in the context of analogies such as those that are normally used in logical [and conceptual] analysis".
19 Many authors also underline the specific differences between the early view, and the full development of dialectic as documented in the *Phenomenology of Spirit*. Pöggeler (1981, 42–45) stresses that Hegel "bases [his fragment about *Glauben und Sein*] on Kant's dialectic in the *Critique of Practical Reason* without adopting the expression 'dialectic' [...] Hegel had his reasons if he reclaimed the title 'dialectic' for his thought only in the second half of the Jena's period". Düsing 1976 reconstructs Hegel's dialectical method from the point of view of the history of its development (*Entwicklungsgeschichte*), showing affinities and differences between the early conception of antinomy and *Vereinigung* (1797 to 1800), the writings and fragments of the Jena's period (1801–1806), and the *Science of Logic* (1812–1816). For a comprehensive account of the differences between the conceptions and uses of dialectic in the different phases of Hegel's thought see Schäfer 2001. Berti 2015 reconstructs the development of Hegel's view accentuating, as I do, its continuity.

of contradiction in his mature writings. In the *Differenzschrift* we also find Hegel's own attempt at formalising *Aufhebung*. Here Hegel writes that:

> If we reflect on the merely formal character of speculation, fixing the synthesis of knowledge analytically, then the antinomy, the contradiction that overcomes and maintains itself (*der sich selbst aufhebende Widerspruch*) is the highest expression of knowledge and of truth. (Hegel Werke 2, 39)

Focusing on "the merely formal character of speculation" and "fixing the synthesis [...] analytically" means considering the form of this peculiar (speculative) content, giving an account of its very structure or form. The result of such a consideration is the following insight, which I call C or "Principle of Contradiction":

C: the contradiction that overcomes and maintains itself is the form of truth.[20]

The line of thought that brings Hegel to state this principle is a reflection on the logical principle of identity. The mere identity $A = A$ ("A is identical to A") corresponds, for Hegel, to the law followed by intellectual thought in its attempts to grasp reality. "$A = A \& A \neq A$" ("A is identical to A and A is not identical to A") is, in contrast, the complete account of the form of truthful thought grasped by reason. The law of identity, according to Hegel, is the result of an abstraction. When we fix this logical and metaphysical principle (this commonly acknowledged law of truth) by saying "everything is equal to itself", we abstract from the fact that it is not only the case that everything is equal to itself. Every thing grasped by thought is equal to itself but it is also equal to something different from itself. For instance, we say "the rose is the rose", but a truthful account of what it means to be a rose, or of what roses are, requires completeness, i.e. taking into account all the properties of the object. Seeing that a rose has many predicates, i.e. is thorny, green, red etc. further implies that the mere identity "a rose is a rose" is not sufficient to give a full account of what it means to be a rose. Hence, as Hegel explains, a full account of a concept necessarily involves overcoming the identity principle, stating not only $A = A$ but also, at a preliminary level: $A = B, C, D$ etc. In the *Differenzschrift* Hegel writes that:

---

[20] This and the following principle S echo the first of Hegel's *Habilitationsthesen* (1801), see Hegel Werke 2, 533: *contradictio est regula veri, non contradictio falsi*.

> [T]he understanding sees in the statements A = B, A = C etc. just a repetition of the A, that is, it holds fast only to the identity and abstracts away from the fact that, repeating A as B or in B means affirming [...] not-A, and that as A, so A as not-A. (Hegel Werke 2, 39)

Reason sees, in contrast, the whole movement from A = A to A = B, C, D to A = Not-A, and comes to the conclusion "A = A and A = Not-A" (which, for Hegel, is one and the same as A = ¬A or A ≠ A). This is what motivates the idea that the LNC is inadequate if we want to give an account of the form of rational thought, that is: a complete thought about what something truly is.

In the *Essay on Scepticism and Philosophy* Hegel focuses on what he calls here "philosophical" or "rational concepts" or "sentences" (*Vernunfterkenntnisse* or *Vernunftsätze*). He explains that the principle of ancient scepticism (I will call it S):

> S: *panti logo logos isos antikeitai* (for every valid argument there is an opposite one that is equally valid)

is implicitly present in every philosophy, and in every philosophical thesis. But why is the sceptical principle a fundamental philosophical principle? Hegel himself considers an example that I have already mentioned: Spinoza's definition of substance as *causa sui*, and as that whose essence implies existence. The sentences expressing such (philosophical, rational, speculative) contents ("the substance is *causa sui*", or "*causa sui* is that whose essence implies existence", or "the substance is that whose essence implies existence"), once we make their contents explicit, entail a link between contradictories (terms and sentences). Hegel thus writes:

> [W]hen in a sentence that expresses a rational content we focus our attention [...] on the concepts that it contains, and on the way in which they are linked to each other, then it turns out [...] that *they are joined in a way according to which they contradict each other* [my emphasis]. (Hegel Werke 2, 229)

"Focusing attention on the concepts" corresponds to considering what the terms joined in a sentence mean. "Focusing attention on the way they are linked with each other" means adopting a formal point of view. From this perspective the sentence "the substance is *causa sui*", once we make the content of the terms *causa* and *sui* and the way they are connected explicit, appears to be a conjunction of contradictories. In the passage considered above from the *Differenzschrift* Hegel stresses that C is the result of applying an intellectual, merely formal consideration to the rational content. Here, he underlines that S is the result of focusing on the *form of speculative contents*. To stick to the example of speculative rational content given in the *Skeptizismusaufsatz*, for something to be a cause means that it cannot be cause of itself, and that it has to be cause of something else. Thus the term "sui" turns out to be the negation of the term "causa", and the sentence entails a contradiction, namely "the substance is cause and the substance is not cause". Similarly, if we make the content of the sentence "the substance is that whose essence entails existence" explicit we have that its terms (essence and existence) are "joined in a way according to which they contradict each other". In other words, saying "essence entails existence" means affirming of essence what is excluded by its very definition (the determination "existence"), and affirming of existence what is excluded by its very definition (the determination "essence").

Thus, in order to give an account of the logic of this kind of rational or philosophical contents we need to dismiss the LNC, and to adopt the sceptical principle S.

> The so called Law of Non-Contradiction has no formal value for reason, so that every sentence of reason, given [the meaning of] its concepts, must entail a violation of it; that a sentence is merely formal means for reason: affirming it alone, without affirming at the same time its contradictory opposite, is false [...] every true philosophy contains this negative part, this eternal violation of the LNC, so, who wants, can single out this negative part and make a scepticism out of everything. (Hegel Werke 2, 230)

It is important to note that Hegel here is not criticising formality *tout court*. In other words, he is not denying the rights of a consideration about the general structure of sentences and arguments. Quite the opposite, he is rather *fixing* the general structure of a particular kind of sentences and arguments, namely the philosophical, rational, and sceptical ones. At the same time he is pointing out that the word "formal", as it is traditionally used, refers to the form of the understanding (LNC and related principles), which, from the point of view of reason, are incomplete, and thus untrue. We have truth, for Hegel, only if we have complete and sceptical thoughts, that is: rational and conceptual thought. The laws given by *Verstandeslogik* are, in this respect, laws of intellectual, i.e. partial and dogmatic (non-critical) thought.

Hegel's considerations on the same topic in the *Wissenschaft der Logik* and in the logic of the *Encyclopaedia* substantially confirm these insights. In the Encyclopaedia § 115 we read that the Principle of Identity (whose negative version is the LNC) and LEM, called by Hegel *Satz des Gegensatzes* ("principle of opposition") belong to what are generally acknowledged as the universal laws of truth. Its positive formulation is:

I: "Everything is identical to itself, $A = A$"

and its negative one

LNC: "A cannot be both A and non-A at the same time". (Hegel Werke 8, 237/ Hegel 1991, 180)

According to Hegel "the very sentential form contradicts them". As a matter of fact, the form of the sentence is, as we have seen, "the S(ubject) is P(redicate)",[21] and implies that something (the grammatical subject) is said to be something else (the grammatical predicate) or that something (the predicate) is said of something else (the subject). The form of the sentence thus could be more adequately expressed by A = B (a rose is a flower), and contradicts I, which states that A = A.

But in order to think, speak, exist or even imagine things we have to break this rule: "speaking in accordance with this supposed law of truth (a planet is – a planet, magnetism is – magnetism, the spirit is – spirit) is rightly regarded as silly" (Hegel Werke 8, 236/Hegel 1991, 180).

In the context of his discussion of the *Reflexionsbestimmungen des Wesens* ("determinations of reflection") in both the *Wissenschaft der Logik* and the *Encyclopaedia* Hegel also criticises the Law of Excluded Middle (LEM). In the *Encyclopaedia* §§ 119 f. Hegel formulates LEM as follows:

LEM: "of two opposed predicates only one comes up to something, and there is no third"

LEM is said to be, exactly like LNC and I, a principle of the abstract understanding. It has three limits. First of all, it is a sign of the "thoughtlessness" of ordinary logic, which puts LEM beside I, without noticing that LEM is the very confutation of I. While I negates that A can be said to be something other than itself, LEM explicitly states it.

Second, LEM entails its own refutation:

The principle of the excluded third is the principle of the determinate understanding, which wants to avoid contradiction, but in so doing falls into it. A must be either + A or – A, thus the third, the A which is neither + nor – and which is thus posited also as both + A and – A, is already expressed. (Hegel Werke 8, 244/Hegel 1991, 185)

---

[21] On Hegel's view on the meaning of the copula see Hegel Werke 8, 323 f./Hegel 1991, 249 f. On the fact that Hegel here does not confound between the "is" of predication and the "is" of identity see here Part III.

In order to state LEM, i.e. that A is either + A or − A we have to presuppose the third, namely "the *A which is neither + nor − and thus both*" (Hegel Werke 8, 244/ Hegel 1991, 185).

Third, LEM declares a concept

> to which neither or both of two mutually contradictory characteristics apply [...] logically false, like for instance, a square circle. Now, although a polygonal circle or a rectilinear arc contradicts this principle just as much, geometers do not hesitate to consider and to treat the circle as polygon with rectilinear sides. (Hegel Werke 8, 245/Hegel 1991, 186)

Another counterexample of LEM is the very concept of circle, which includes as its essential characteristics the determinations of centre and periphery, characteristics that, according to Hegel, are opposed to each other and contradict each other (Hegel Werke 8, 245/Hegel 1991, 186). Evidently, LNC and LEM are truly challenged only if the predicates or sentences at stake are contradictory, and one could still wonder why different determinations such as "center" and "periphery" have to be considered as contradictory predicates, predicates one of which is the negation of the other. The reason relates, Hegel writes, to the kind of thought practiced by philosophers:

> Ordinary consciousness treats the distinct terms as indifferent to one another. Thus we say, 'I am a human being, and I am surrounded by air, water, animals and everything else.' In this ordinary consciousness everything falls outside everything else. The purpose of philosophy is, in contrast, to banish indifference and to know the necessity of things, so that the other is seen to confront *its* other [i.e. its negation] [...] true thought is the thought of necessity. (Hegel Werke 8, 246/Hegel 1991, 187)

From this point of view, it is clear that a more suitable principle than LEM for expressing the form of rational, philosophical thought is what we can call the Opposition Principle:

O: "Everything stands in opposition"

Nothing, says Hegel, exhibits the abstract "either-or" of the abstract intellect.

> Everything that is at all is concrete, and hence it is inwardly distinguished and self-opposed [...] generally speaking is contradiction that moves the world, and it is ridiculous to say that contradiction cannot be thought. What is correct in this assertion is just that contradiction is not all there is to it, and contradiction sublates itself by its own doing. Sublated contradiction, however, is not abstract identity, for that is itself only one side of the opposition. (Hegel Werke 8, 246f./Hegel 1991, 187)

"Contradiction is not all there is to it" means that not everything is contradictory, for Hegel. "Contradiction sublates/overcomes itself" means that the conflictual situations in which we cannot admit a without deriving from it $\neg a$ and vice versa are indeed solved. However, they are not solved in that we only admit a, or only $\neg a$ (abstract identity) as true, but rather insofar as we admit that the biconditional $a \leftrightarrow \neg a$ is true.

In the *Science of Logic* Hegel points out the differences between LEM and LNC:

> [T]he law of the excluded middle is also distinguished from the laws of identity and contradiction considered above; the latter of these asserted that there is nothing that is at once A and not-A. [LEM] implies that there is nothing that is neither A nor not-A, that there is not a third that is indifferent to the opposition. (Hegel Werke 6, 74/Hegel 1969, 438f.)

Hegel stresses here, like he does in the *Encyclopaedia*, that in fact "the third" that is indifferent to the opposition is given in the law itself ("A itself is present in it"), and he claims that the third, which has here (as "neither nor") "the form of a dead something, when taken more profoundly" is nothing else than the unity of the opposites (Hegel Werke 6, 74/Hegel 1969, 438f.). Thus Hegel is here suggesting that there is a third, and that the third is not the "neither nor" (the "dead something") but rather "both" (the contradiction).[22]

---

[22] See also Hegel Werke 19, 399/Hegel 1892ff., vol. II., 369.

16 Hegelian paraconsistentism

In the preceding chapters I have isolated some logical peculiarities concerning the meaning of negation in dialectic, and of the connective joining the two terms of a dialectical contradiction.

In short, the logical constraints connected to the speculative dialectical approach are:

a) Hegelian negation has three features: it is internal determination of concepts, it has positive partial content, and its iteration does not produce affirmation but contradiction (see the law DDN fixed above).

b) Hegelian *Vereinigung* stands for the link that joins the two terms of a true contradiction. It is not simplifiable: the two terms, separately taken, are untrue since they only convey partial truth. Logically Hegel's *Vereinigung* corresponds to a true biconditional of the form $a \leftrightarrow \neg a$.

c) All these features do not mark a special conception of contradiction, negation, and conjunction. In light of the view that true contradictions for Hegel are contradictions of the form $a \leftrightarrow \neg a$ all the features (failure of simplification, negation's partiality and positivity, DDN) are perfectly understandable in classical terms. The only dissonance with the classical paradigm is Hegel's admission that contradictions can be true, and more precisely Hegel's law C, according to which contradiction is the form of truth. This aspect marks the vicinity of Hegel's logic to paraconsistent logics. Yet it also marks its difference from extant dialetheic and paraconsistent approaches.

d) The biconditional $a \leftrightarrow \neg a$ is for Hegel the form of conceptual, philosophical and speculative thought, it is the form fixed by *Vernunftlogik*, different from the traditional laws of truth fixed by *Verstandeslogik* (LNC, I and LEM). While contradiction (as *Vereinigung*) is the form (norm, and law) of truth (see the principle C isolated above), LNC and LEM are the forms of correctness (which I have called R), or partial truth.

Paraconsistent logic, and more particularly dialetheism, the perspective according to which there are true contradictions, is a crucial reference point in order to understand the logical relevance of Hegel's thoughts on true contradictions. From a dialetheic point of view, paradoxical deductions convey a true biconditional of the form $a \leftrightarrow \neg a$. I have argued (see IV) that the same structure is typical of dialectical deductions, even if Hegel's understanding of dialectical deductions, as we have seen, questions some assumptions shared by dialetheists, e.g. that CM holds.

All this shows that Hegel's theory of contradiction, and of its link to truth, belongs to the family of theories according to which (some) contradictions are admissible, but also presents elements that are not contemplated by any extant paraconsistent semantics. More precisely, from the perspective of what one could call *Hegelian paraconsistentism* a crucial problem at the core of dialetheic semantics of contradictions can be addressed, namely what I call the *criterion problem*.

The fundamental question for philosophers who hold that there are true contradictions is giving a criterion according to which it is possible to distinguish between true and false, admissible and non-admissible contradictions, and to block the so called explosion. One of the most important proponents of dialetheism, Graham Priest, observes:

> I am frequently asked for a criterion as to when contradictions are acceptable and when they are not. It would be nice if there were a substantial answer to this question [...] But I doubt that this is possible [...] One can determine the acceptability of any given contradiction, as of anything else, only on its individual merits. (Priest 2004, 35)

The refusal to give a criterion for distinguishing between true and false contradictions is seen by many as a major problem, which renders it impossible for a dialetheist to disagree, and to argue consistently for her position.[23]

In short, dialetheists hold that from the biconditional $a \leftrightarrow \neg a$ we can derive the contradiction $a \wedge \neg a$, which is true, and whose conjuncts are both true and false, and can be stated separately.[24] Hence, from a dialetheic point of view the inference

$$a \leftrightarrow \neg a \vdash a$$

is valid. *In contrast, for Hegel, the same inference is not valid.* This marks a major difference between Hegelian and dialetheic paraconsistentism. In fact, when Hegel claims that "*Vereinigung* is the norm" he factually emphasises a basic behaviour of the contradictory biconditional $a \leftrightarrow \neg a$. If $a \leftrightarrow \neg a$ is true, then the conjunction $(a \rightarrow \neg a) \wedge (\neg a \rightarrow a)$ is true, but $a \rightarrow \neg a$ or $\neg a \rightarrow a$, taken separately, are not. Consequently, from a Hegelian perspective we cannot infer $\neg a$ from $a \rightarrow \neg a$, and a from $\neg a \rightarrow a$.

---

23 See Berto 2006, 283–297, Shapiro 2004, 336–354, Littmann/Simmons 2004, 314–335.
24 See Priest 2006, 9 ff. On the dialetheic argument from the equivalence $a \leftrightarrow \neg a$ to the contradiction $a \wedge \neg a$ see Field 2008, 7 and d'Agostini 2009, 151 ff.

In my interpretation Hegel's perspective on contradictions implies, in other words, that it would be wrong to interpret $a \leftrightarrow \neg a$ as a biconditional in which the sides from left to right: $a \rightarrow \neg a$, and from right to left: $\neg a \rightarrow a$, are separable.

Let us go back to the example considered in 8.2., the double sentence "what is rational is real and what is real is rational" which I have expressed formally as a biconditional of the form $\forall x\, (RAx \leftrightarrow REx)$. If we simplify the conjunction conveyed by the biconditional: $(RAx \rightarrow REx) \wedge (REx \rightarrow RAx)$ stating only the side from right to left (what is real is rational), we simply misread Hegel's idea, interpreting it as a mere legitimation of what is there (in the specific case: the Prussian state of Hegel's times). If we simplify the conjunction stating the sole side from left to right of the biconditional (what is rational is real), we misread, again, Hegel's claim, reducing it, for example, to the view that every rational insight is actual, and what is irrational simply does not exist (it is as if we said that racism does not exist because it is irrational, or that, since global justice is rational, there is justice everywhere in the world). The same can be seen considering other conceptual pairs examined by Hegel: being and nothing, essence and existence, being in consciousness and being out of consciousness. Clearly, separating the conceptual pairs of a Hegelian contradictory concept produces irretrievable mistakes.

All this explains in what sense we can hold that Hegelian contradictions are special biconditionals. From Hegelian contradictions $(a \leftrightarrow \neg a)$ we cannot derive usual conjunctive contradictions $(a \wedge \neg a)$.

A final question needs to be addressed: is my formal account of Hegelian contradictions of any relevance with respect to the crucial question at the core of every paraconsistent logic, namely the necessity to avoid explosion, Apostel 1979 recalls that paraconsistent logics, which were impressively growing in the 70ies, and were developed by the da Costa school in Brasil, by Jaskowski in Poland and by Routley in Australia, present the necessary condition and the formal basis of dialectics. However, he also claims that they cannot be said to be dialectical logics in the Hegelian sense, and, more importantly, that they need dialectical logic. They allow us to see how to logically deal with contradictions without explosion, but they do not let us see why and how we can affirm a contradiction. Hence Apostel (1979, 459) formulates the following task for a dialectical foundation of paraconsistentism: "in dialectical logic we have to show what contradictions are admissible and what are not". Yet Apostel does not give a criterion to establish what are the admissible contradictions.

In this scenario, Hegel's approach to contradictions is a genuine answer to Apostel's request, and thus a solution to the criterion problem. The idea of Hegelian contradictions as true biconditionals of the form $a \leftrightarrow \neg a$ explains why Hegel's logic is not explosive and why Hegel was not a trivialist. Moreover, it constitutes an option for contemporary philosophers who are interested in distinguishing true from false, admissible from non-admissible contradictions.

From a Hegelian perspective not every contradiction $a \wedge \neg a$, but only those contradictions that manifest the necessary inferential link between a and $\neg a$ expressed by the biconditional $a \leftrightarrow \neg a$ are true. Hegel admits that from the inferences "if the one is one then it is many", and "if the one is many then it is one", we can derive that the concept of the one contains both properties: being one and being many. This seems to suggest that Hegel admits the classical derivation of contradiction from equivalence. However, his conjuncton of contradictories has truth-conditions that clash against the ones normally presented by conjunctions. From the Hegelian true contradiction "a and not-a" we cannot infer only one of the two conjuncts. The behaviour of the conjunction expression of a Hegelian true contradiction, in particular the failure of simplification, does not mean that negation does not work classically, or that the conjunction Hegel is writing about is non-classical. It simply means that, when Hegel talks about contradictions in conjunctive terms as "a and not-a" he refers to the structure conveyed by the biconditional contradiction $a \leftrightarrow \neg a$.

In this light, we can fix a criterion in order to distinguish true from false contradictions, solving the dialetheic criterion problem, and answering Apostel's request. Only those contradictions whose terms are not separable, and which manifest the form $a \leftrightarrow \neg a$, are true, and do not produce explosion. Those contradictions whose terms are separable are not true contradictions, and generate explosion.

Let us see the failure of explosion in the case of Hegelian true contradictions in more detail. The term "explosion" is used today with reference to the law known since the Middle Ages as *ex contradictione* (or *ex falso*) *sequitur quodlibet* (ECQ). The law is also called *Pseudo-Scotus Law* because it was discussed in a text wrongly attributed to Duns Scotus (*In universam logicam quaestiones*).[25] ECQ states that from a contradiction everything follows. Its proof in the original Latin formulation is:

ECQ
1. *Sortes est et Sortes non est*
2. *Sortes est*
3. *Sortes est vel homo est asinus*
4. *Sortes non est*

---

[25] For a detailed discussion of ECQ, its history, its possible formal expressions, its proofs, as well as the proof's possible critiques see Berto 2007a, 107 ff.

5. Homo est asinus

The proof shows that from the contradiction *Sortes est et Sortes non est* we can logically derive any sentence. The admission of one contradiction produces explosion, that is trivialism (the view that everything is true) and the loss of every logical constraint. The passages from 1. to 5. are all motivated by classical laws: *simplification* explains the passage from 1. to 2.; *disjunction introduction* the one from 2. to 3.; in the steps from 3. to 5. are used *simplification* (from 1.) and *disjunctive syllogism* (from 3. and 4.).

Standardly, paraconsistent logicians try to prevent explosion by working on the admissibility of disjunctive syllogism (the passage from 3. and 4. to 5.). The dismissal of disjunctive syllogism is highly controversial, even among paraconsistent logicians.[26] The Hegelian perspective on admissible ($a \leftrightarrow \neg a$) versus non-admissible ($a \wedge \neg a$) contradictions allows for a way to stop explosion in a much less demanding way. Admitting that a true contradiction has the form $a \leftrightarrow \neg a$, whereby the two conjuncts $a \to \neg a$ and $\neg a \to a$ are inseparable, means that we cannot even state the two conjuncts a and $\neg a$ of a conjunctive contradiction $a \wedge \neg a$ separately, and that the step from 1. to 2. is invalid. Hence the argument for explosion, in cases of Hegelian contradictions, fails.

---

[26] For an overview on the different paraconsistent critiques of the proof for *ex contradictione quodlibet* see Berto 2007a, 111 ff. Relevant logicians have developed the most detailed arguments against the proof based on the rejection of disjunctive syllogism. For a reconstruction of these arguments as well as some of their problematic implications see Bremer 1998, 53 ff. and 69 ff. as well as Berto 2007a, 114 and 187 ff.

Summary

The fifth part is a consideration of dialectical contradictions. I examine here what Hegel says on the elements constituting a contradiction, i.e. what Hegel says on the connective uniting the two poles of an antinomy, as well as Hegel's theses on the negation expressing the antinomic relation between them.

Chapter 13 concerns the concept of *Vereinigung*, used by the young Hegel in the fragment on *Glauben und Sein* 1797–98 (before reappearing in the *Science of Logic*) in order to express the connective joining the two elements of an antinomy. Hegel's theory is that "*Vereinigung* [unification] is the norm". In other words, the union of "matter is continuous" and "matter is not continuous" or "human beings are free" and "human beings are not free" completely and adequately expresses the concept (of matter, and of human freedom), while the single sentences forming the antinomy, taken in isolation, are insufficient to express the complex and heterogeneous nature of concepts. I suggest how in Hegel's theory of *Vereinigung* a first problem concerning dialectical contradictions, which I also consider in the chapter on negation, is immediately solved: Hegelian contradictions are antinomies, i.e. they can be expressed as couples of sentences one of which is the negation of the other. They can be considered in all respects as contradictions in the standard logical meaning of the word. I also claim here that, from a logical point of view, Hegelian *Vereinigung* can be expressed as a biconditional rather than a simple conjunction. The link established by *Vereinigung* is stronger than the one conveyed by a simple conjunction. In it the two antinomic theses are not linked paratactically, but rather *necessitate* each other.

In **Chapter 14** I examine Hegel's conception of negation. Synthetically, some aspects of Hegel's conception conform to the semantics of negation in classical logic. I explain that for Hegel the opposition between predicates is expressed syntactically in terms of a relation between sentences one of which is the negation of the other. Hegel's oppositions as couples of contraries (for example matter's "being continuous", P, and its "being punctual", Q) are turned into couples of contradictory predicates (its "being continuous", P, and its "not being continuous", $\neg P$). The relation between contradictory predicates is finally expressed sententially: "matter is continuous" (p) and "matter is not continuous" ($\neg p$). Moreover, negation for Hegel is to be intended as a contradictory forming operator (whereby contradictions, for Hegel, as it is showed in the last chapter, are not explosive). In this respect, what Hegel intends by "negation" is pretty close to what logicians understand by this operator. The peculiarly Hegelian traits, which mark the difference between Hegel's notion and the classical one, are: negation for Hegel is a) *internal* to the conceptual content; it is b) *pos-*

itive, and c) the iteration of negation does not produce affirmation (as it happens in the case of classical double negation), but rather contradiction. My thesis is that to understand these peculiarities one should stick to Hegel's idea of "*Vereinigung* is the norm". In other words, all these peculiarities are grounded in and explainable in light of *Vereiningung* as a biconditional of the form $a \leftrightarrow \neg a$.

a) That negation is internal means that it is within the biconditional, and not outside of it. If the fully developed concept of human freedom implies the double conditional "human beings are free in their necessity and subjected to necessity in their freedom" or, which for Hegel is the same, "human beings are free if not free, and not free if free", then "human beings are not free" stays within, is a part or property internal to the concept of freedom, and not the negation of the concept of freedom. b) That the negation is positive means that $\neg a$ is not cancellation of a. From the point of view of the conceptual content expressed in terms of $a \leftrightarrow \neg a$, it is evident that affirming $\neg a$ does not mean to cancel a, because the peculiar nature of conceptual contents is such that, in stating $\neg a$, we reproduce a, since from $\neg a$ we derive a. c) Finally, the double negation of a does not correspond to affirming a because, from the point of view of the expression of conceptual truth in terms of $a \leftrightarrow \neg a$, *it is not the case that human beings are not free* does not mean that human beings are free, but rather that they are free if and only if they are not free.

In **Chapter 15** I examine Hegel's treatment of the laws of intellectual logical (*verstandeslogische*) truth, namely, LNC and LEM. For Hegel they are not laws of rational truth (*Wahrheit*) but rather of mere *Richtigkeit*. In other words, Hegel explicitly states that "the highest expression", "the norm", and "the formal character" of speculation (or rational or philosophical thought) is contradiction. Rational or philosophical truth includes scepticism: the thesis according to which *for every logos (valid argument/apparently true sentence) there is an opposite one which is equally valid/true* is the very condition for thinking validly and truthfully. Hence rational logic entails a critique of LNC and LEM. At the same time, Hegel claims that the negation of LNC and the view according to which contradiction is the form/norm of truth are the negative part of philosophy, and that "contradiction is not everything". That is, as we have seen in Part IV, Hegel does not think that everything is contradictory. He holds that the contradiction expression of the philosophical true is only true, and not also false. Hence philosophy and rational thought for Hegel also has a positive, dogmatic side, and not only a negative or sceptical one.

In the last chapter (**Chapter 16**) I analyse Hegel's position on contradictions by confronting it with the paraconsistent (and in particular dialetheist) theories according to which admitting true contradictions does not imply the explosion of the language in which they are formulated, and of logic. The consideration of Hegel's account of *Vereinigung*, negation, and LNC shows that Hegel can, in all respects, be considered a paraconsistent philosopher, since, as I show at the end of the chapter, in dialectical logic the admission of true contradictions does not imply explosion. At the same time Hegel's dialectical logic is different from the dominant options in discussions on true contradictions and dialetheism. More specifically, as I have highlighted in Part IV, dialectical inferences share with paradoxical deductions the derivation of a true biconditional of the form $a \leftrightarrow \neg a$. While from the dialetheic point of view it is possible to draw from $a \leftrightarrow \neg a$ the conjunction $a \wedge \neg a$ (and hence, through the principle of simplification, only a, or only $\neg a$), I argue that this inference, in principle, does not adequately express the dialectical point of view. Hegelian contradictions are paradoxical biconditionals of the form $a \leftrightarrow \neg a$ but, from a dialectical point of view, it would be a mistake to infer from $a \leftrightarrow \neg a$ only $a \rightarrow \neg a$ or only $\neg a \rightarrow a$. Hence, from a dialectical point of view the inference of $\neg a$ from $a \rightarrow \neg a$ or of a from $\neg a \rightarrow a$ should not be admissible. For example, Hegel's concept of freedom implies that humans are free if subjected to necessity and subjected to necessity if free. Inferring from this only that "if humans are free then they are subjected to necessity" would be wrong, as it would imply, for example, the total delegitimation of the concept of freedom. Conversely, only holding that if humans are subjected to necessity then they are free would imply, for example, the view that coercive instruments such as tortures, imprisonments etc. are always perfectly acceptable, since they would be used in the name of human freedom. In conclusion, I show that the relation of reciprocal foundation between $\neg a$ and a expressed by Hegel's concept of contradiction is useful for solving a problem at the heart of dialetheism (and, more generally, fundamental for philosophically grounding paraconsistentism). It is what I call the *criterion problem*, the absence of criteria in order to distinguish between admissible and non-admissible contradictions. I state that dialectical contradictions have the form of a paradoxical biconditional $a \leftrightarrow \neg a$, whose conjuncts $a \rightarrow \neg a$ and $\neg a \rightarrow a$ cannot be simplified, and which cannot be turned into a paratactic contradiction $(a \wedge \neg a)$. Dialectical contradictions, so understood, are the only admissible ones. On this basis, it is easy to show how the principle called *explosion* (or *ex contradictione quodlibet*) does not hold for dialectical contradictions.

Bibliography

Abel, Günter (1999): *Sprache, Zeichen, Interpretation*. Frankfurt a. M.: Suhrkamp.
Abel, Günter (2004): *Zeichen der Wirklichkeit*. Frankfurt a. M.: Suhrkamp.
Adorno, Theodor Wiesengrund (1966): *Negative Dialektik*. Frankfurt a. M.: Suhrkamp.
Adorno, Theodor Wiesengrund (2010): *Einführung in die Dialektik*. Frankfurt a. M.: Suhrkamp.
Alston, William (1996): *A Realist Conception of Truth*. Ithaca and London: Cornell University Press.
Ameriks, Karl/Stolzenberg, Jürgen (eds.) (2005): *Internationales Jahrbuch des Deutschen Idealismus/International Yearbook of German Idealism: German Idealism and Contemporary Analytic Philosophy*. Vol. 3. Berlin New York: De Gruyter.
Apostel, Leo (1979): "Logica e dialettica in Hegel". In: Marconi, Diego (ed.): *La formalizzazione della dialettica*. Torino: Rosenberg & Sellier, 5–113.
Armstrong, David Malet (2010): *Sketch for a Systematic Metaphysics*. Oxford: Oxford University Press.
Asmus, Conrad/Restall Greg (2012): "History of the Consequence Relation". In: Gabbay, Dov/Pelletier, Francis/Woods, John (eds.): *Handbook of the History of Logic*. Amsterdam: Elsevier, 11–62.
Asmuth, Christoph (2006): *Interpretation – Transformation. Das Platonbild bei Fichte, Schelling, Hegel, Schleiermacher und Schopenhauer und das Legitimationsproblem der Philosophiegeschichte*. Göttingen: Vandenhoeck & Ruprecht.
Asmuth, Christoph (2007): "Realismus und Idealismus oder: Werden philosophische Probleme entschieden?". In: *Prolegomena* 7/2, 203–221.
Aubenque, Pierre (1990): "Hegel und die Aristotelische Dialektik". In: Riedel, Manfred (ed.): *Hegel und die antike Dialektik*. Frankfurt a. M.: Suhrkamp, 208–226.
Baldwin, Thomas (1991): "The Identity Theory of Truth". In: *Mind* 100. No. 397, 35–52.
Baldwin, Thomas (2004): "Über Wahrheit und Identität". In: Halbig, Christoph/Quante, Michael/Siep, Ludwig (eds.): *Hegels Erbe*. Frankfurt a. M.: Suhrkamp, 21–45.
Baptist, Gabriella (1986): "Hegel e l'*Antitetica* della *Critica della ragion pura*". In: *Paradigmi* IV. No. 11, 271–297.
Barnes, Johnathan (ed.) (1984): *The Complete Works of Aristotle*. Princeton: Princeton University Press.
Barone, Francesco (1957): *Logica formale e logica trascendentale*. Torino: Edizioni di Filosofia.
Baum, Manfred (1986): "Anmerkungen zum Verhältnis von Systematik und Dialektik bei Hegel". In: Henrich, Dieter (ed.): *Hegels Wissenschaft der Logik*, Stuttgart: Klett-Cotta, 65–76.
Baum, Manfred (1990): "Kosmologie und Dialektik bei Platon und Hegel". In: Riedel, Manfred (ed.): *Hegel und die antike Dialektik*. Frankfurt a. M.: Suhrkamp, 192–207.
Baumann, Charlotte (2019): "Hegel's realm of shadows: logic as metaphysics in the science of logic". In: *British Journal for the History of Philosophy* 27. No. 6, 1256–1260.
Baumgarten, Alexander (1739): *Metaphysica*. Halle: Hemmerde.
Beall, Jc/Restall, Greg (2014): "Logical Consequence". In: Zalta, Ed (ed.): *The Stanford Encyclopedia of Philosophy*. http://plato.stanford.edu/archives/fall2014/entries/logical-consequence/, accessed 05/06/2020.
Becker, Werner/Essler, Wilhelm (eds.) (1981): *Konzepte der Dialektik*. Frankfurt a. M.: Klostermann.

Beiser, Frederick (1993): "Hegel and the Problem of Metaphysics". In: Beiser, Frederick (ed.): *The Cambridge Companion to Hegel*. Cambridge: Cambridge University Press, 1–24.
Beiser, Frederick (2005): *Hegel*. London: Routledge.
Beiser, Frederick (2008): "The Puzzling Hegel-Renaissance". In: Beiser, Frederick (ed.): *The Cambridge Companion to Hegel and Nineteenth Century Philosophy*. Cambridge: Cambridge University Press, 1–14.
Bellissima, Fabio/Pagli, Paolo (1996): *Consequentia Mirabilis. Una regola logica tra matematica e filosofia*. Firenze: Olschki.
Bencivenga, Ermanno (2000): *Hegel's Dialectical Logic*, Oxford: Oxford University Press.
Berti, Enrico (1990): "Hegel und Parmenides oder: Warum es bei Parmenides keine Dialektik gibt". In: Riedel, Manfred (ed.): *Hegel und die antike Dialektik*. Frankfurt a. M.: Suhrkamp, 65–83.
Berti, Enrico (2015): *Contraddizione e dialettica negli antichi e nei moderni*, Brescia: Morcelliana (first edition 1987).
Bertinetto, Alessandro (ed.) (2004): *J. G. Fichte. Logica trascendentale II*. Napoli: Guerini 2004.
Berto, Franz (2005): *Che cosa è la dialettica Hegeliana?*. Padova: Il Poligrafo.
Berto, Franz (2006): "Meaning, Metaphysics, and Contradictions". In *American Philosophical Quarterly* 43. No 4, 283–297.
Berto, Franz (2007a): *How to Sell a Contradiction. The Logic and Metaphysics of Inconsistency*. London: College Publications.
Berto, Franz (2007b): "Hegel's dialectics as a semantic theory. An analytic reading". In: *European Journal of Philosophy* 15. No 1, 19–39.
Bochenski, Joseph Maria (1978): *Formale Logik*. Freiburg München: Alber.
Bodammer, Theodor (1969): *Hegels Deutung der Sprache*. Hamburg: Meiner.
Boeder, Heribert (1990): "Vom Schein einer Heraklitischen Dialektik". In: Riedel, Manfred (ed.): *Hegel und die antike Dialektik*. Frankfurt a. M.: Suhrkamp, 98–108.
Boole, George (1847): *The mathematical analysis of logic: being an essay towards a calculus of deductive reasoning*. Cambridge: Macmillian.
Bordignon, Michela (2013): "Dialectic and Natural Language. Theories of Vagueness". In: *Teoria* 33. No 1, 179–198.
Bordignon, Michela (2014): *Ai limiti della verità. Il problema della contraddizione nella logica di Hegel*, Pisa: ETS.
Boyer, Carl (1968): *A History of Mathematics*. New York London Sydney: John Wiley & Sons.
Brandom, Robert (2002): *Tales of the Mighty Dead*. Cambridge MA: Harvard University Press.
Brandom, Robert (2005): "Sketch of a Program for a Critical Reading of Hegel. Comparing Empirical and Logical Concepts". In: *Internationales Jahrbuch des Deutschen Idealismus* 3, 131–161.
Brandom, Robert (2014): "Some Hegelian Ideas of Note for Contemporary Analytic Philosophy". In: *Hegel Bulletin* 35. No 1, 1–15.
Breidbach, Olaf (2006): *Goethes Metamorphosenlehre*. München: Fink.
Bremer, Manuel (1998): *Wahre Widersprüche. Einführung in die parakonsistente Logik*, Sankt Augustin: Academia Verlag.
Brinkmann, Klaus (1994): "Hegel's Critique of Kant and Pre-Kantian Metaphysics". In: Pinkard, Terry/Engelhardt, Tristram (eds.): *Hegel Reconsidered*. Dordrecht: Kluver, 57–68.

Bubner, Rüdiger (1990): "Dialektik oder die allgemeine Ironie der Welt. Hegels Sicht des Eleatismus". In: Riedel, Manfred (ed.): *Hegel und die antike Dialektik*. Frankfurt a. M.: Suhrkamp, 84–97.
Buchner, Hartmut (1990): "Skeptizismus und Dialektik". In: Riedel, Manfred (ed.): *Hegel und die antike Dialektik*. Frankfurt a. M.: Suhrkamp, 227–243.
Burbidge, John (2004): "Hegel's Logic". In: Gabbay, Dov/Woods, John (eds.): *Handbook of the History of Logic*. Vol. 3. Bonn: Elsevier, 131–176.
Burgess, John (2009): *Philosophical Logic*. Princeton: Princeton University Press.
Buridan (2014): *Treatise on Consequences*. Translated and with an Introduction by S. Read. New York: Fordham University Press.
Burkhard, Berndt (1993): *Hegels Wissenschaft der Logik im Spannungsfeld der Kritik*. Hildesheim: Olms.
Butler, Clark (1975): "On the Reducibility of Dialectical to Standard Logic". In: *The Personalist* 54. No 4. 414–431.
Butler, Clark (2012): *The Dialectical Method. A Treatise Hegel Never Wrote*. New York: Prometheus Books.
Campogiani, Marco (2006): *Hegel e il linguaggio. Dialogo, lingua, proposizioni*. Napoli: La Città del Sole.
Caron, Maxence (2006): *Etre et identité. Méditations sur la logique de Hegel et sur son essence*. Paris: Passages.
Cassirer, Ernst (1982): "Der Begriff der symbolischen Form im Aufbau der Geisteswissenschaften". In: Ollig, Hans (ed.): *Neukantianismus. Texte*. Stuttgart: Reclam, 127–163.
Chalmers, David/Manley, David/Wasserman, Ryan (eds.) (2009): *Metametaphysics*, Oxford: Oxford University Press.
Cislaghi, Federica (2008): *Goethe e Darwin. La filosofia delle forme viventi*, Milano: Mimesis.
Cobben, Paul/Cruysberghs, Paul/Jonkers, Peter/De Vos, Lou (eds.) (2006): *Hegel Lexikon*, Darmstadt: Wissenschaftliche Buchgesellschaft.
Colombo, Enrico (1998): *Antidialettica. Polemiche sul sistema hegeliano*. Milano: Unicopli.
Cook, Roy (2009): *A Dictionary of Philosophical Logic*. Edinburgh: Edinburgh University Press.
Cortella, Lucio (1995): *Dopo il sapere assoluto*. Milano: Guerini.
Croce, Benedetto (2006): *Saggio sullo Hegel. Seguito da altri scritti di storia della filosofia*, edited by Savorelli, Alessandro and Cesa, Claudio. Napoli: Bibliopolis.
d'Agostini, Franca (2000): *Logica del nichilismo*. Roma Bari: Laterza.
d'Agostini, Franca (2002): *Disavventure della verità*. Torino: Einaudi.
d'Agostini, Franca (2003): "Pensare con la propria testa. Problemi di filosofia del pensiero in Hegel e Frege". In: Vassallo, Nicla (ed.): *La filosofia di Gottlob Frege*. Milano: Franco Angeli, 59–94.
d'Agostini, Franca (2008a): *The Last Fumes. Nihilism and the Nature of Philosophical Concepts*. Aurora: The Davies Group Publishers.
d'Agostini, Franca (2008b): "Metafisica analitica?". In: *Giornale di Metafisica* 30, 243–270.
d'Agostini, Franca (2009): *Paradossi*, Roma: Carocci.
d'Agostini, Franca (2010): "Was Hegel Noneist, Allist or Someist?". In: Nuzzo, Angelica (ed.): *Hegel and the Analytic Tradition*. New York: Continuum, 135–157.
d'Agostini, Franca (2011a): *Introduzione alla verità*. Torino: Bollati Boringhieri.

d'Agostini, Franca (2011b): "Hegel's Interpretation of Megarian Paradoxes. Between Logic and Metaphilosophy". In: Ficara, Elena (ed.): *Die Begründung der Philosophie im Deutschen Idealismus*. Würzburg: Königshausen & Neumann, 121–140.
Dangel, Tobias (2013): *Hegel und die Geistmetaphysik des Aristoteles*. Berlin New York: De Gruyter.
De Boer, Karin (2010): "Hegel's Account of Contradiction in the Science of Logic Reconsidered". In: *Journal of the History of Philosophy* 48. No. 3, 345–373.
De Boer, Karin (2011): "Kant, Hegel, and the System of Pure Reason". In: Ficara, Elena (ed.): *Die Begründung der Philosophie im Deutschen Idealismus*. Würzburg: Königshausen & Neumann, 77–90.
de Laurentiis, Allegra (2002): "Hegel's Concept of Thinking". In: Hüning, Dieter/Stiening, Gideon/Vogel, Ulrich (eds.): *Societas rationis. Festschrift für Burkhard Tuschling zum 65. Geburtstag*. Berlin: Duncker & Humblot, 263–285.
de Laurentiis, Allegra (2005): *Subjects in the Ancient and Modern World. On Hegel's Theory of Subjectivity*. New York: Palgrave Macmillan.
de Laurentiis, Allegra (2016): *Hegel and Metaphysics*. Berlin New York: De Gruyter.
Demmerling, Christoph (1992): "Philosophie als Kritik. Grundprobleme der Dialektik Hegels und das Programm kritischer Theorie". In: Demmerling, Christoph/Kambartel, Friedrich (eds.): *Vernunftkritik nach Hegel*. Frankfurt a.M.: Suhrkamp, 67–99.
De Vos, Lou (1983): *Hegels Wissenschaft der Logik: Die absolute Idee. Einleitung und Kommentar*. Bonn: Bouvier.
De Vos, Lou (2006): "Form". In: Cobben, Paul/Cruysberghs, Paul/Jonkers, Peter/De Vos, Lou (eds.): *Hegel Lexikon*. Darmstadt: Wissenschaftliche Buchgesellschaft, 209–211.
D'Hondt, Jacques (1968): *Hegel en son temps*. Paris: Éditions sociales.
di Giovanni, George (2007): "'Das Logische' of Hegel's Logic in the Context of Reinhold's and Fichte's Late Theories of Knowledge". In Bubner, Rüdiger/Hindrichs, Günnar (eds.): *Von der Logik zur Sprache. Stuttgarter Hegel Kongress 2005*. Stuttgart: Klett-Cotta, 71–87.
Dilthey, Wilhelm (1921): *Die Jugendgeschichte Hegels und andere Abhandlungen zur Geschichte des deutschen Idealismus*. Leipzig: Teubner.
Dottori, Riccardo (2006): *Die Reflexion des Wirklichen*. Tübingen: Mohr Siebeck.
Düsing, Klaus (1976): *Das Problem der Subjektivität in Hegels Logik*. Bonn: Bouvier.
Düsing, Klaus (1983): *Hegel und die Geschichte der Philosophie. Ontologie und Dialektik in Antike und Neuzeit*. Darmstadt: Wissenschaftliche Buchgesellschaft.
Düsing, Klaus (1990): "Formen der Dialektik bei Plato und Hegel". In: Riedel, Manfred (ed.). *Hegel und die antike Dialektik*. Frankfurt a.M.: Suhrkamp, 169–191.
Düsing, Klaus (2012): *Aufhebung der Tradition im dialektischen Denken*. München: Fink.
Dummett, Michael (1991): *Frege and Other Philosophers*, Oxford: Clarendon.
Dummett, Michael (1993): *Origins of Analytical Philosophy* (second edition). London New York: Bloomsbury.
Dutilh Novaes, Catarina (2011): "The Different Ways in which Logic is (said to be) Formal". In: *History and Philosophy of Logic* 32. No. 4, 303–332.
Dutilh Novaes, Catarina (202+): *The Dialogical Roots of Deduction*. Cambridge: Cambridge University Press.
Eckermann, Johann Peter (1987): *Gespräche mit Goethe in den letzten Jahren seines Lebens*. Frankfurt a.M.: Insel.

Emundts, Dina (2018): "Die Lehre vom Wesen. Dritter Abschnitt. Die Wirklichkeit". In: Quante, Michael/Mooren, Nadine (eds.): *Kommentar zu Hegels Wissenschaft der Logik*. Hamburg: Meiner, 387–456.

Emundts, Dina/Sedgwick, Sally (eds.) (2017): *Logik/Logic. Internationales Jahrbuch des Deutschen Idealismus*. Berlin New York: De Gruyter.

Engelhard, Kristina (2007): "Hegel über Kant. Die Einwände gegen den transzendentalen Idealismus". In: Heidemann, Dietmar-Hermann/Krijnen, Christian (Eds.): *Hegel und die Geschichte der Philosophie*, Darmstadt: Wissenschaftliche Buchgesellschaft, 150–170.

Engelhardt, Tristram/Pinkard, Terry (eds.) (1994): *Hegel Reconsidered. Beyond Metaphysics and the Authoritarian State*. Dordrecht: Kluver.

Ferrarin, Alfredo (2001): *Hegel and Aristotle*. Cambridge New York: Cambridge University Press.

Ferrater Mora, José (1963): "On the Early History of Ontology". In: *Philosophy and Phenomenological Research* 24. No. 1, 36–47.

Ficara, Elena (2006): *Die Ontologie in der "Kritik der reinen Vernunft"*. Würzburg: Königshausen & Neumann.

Ficara, Elena (2011): "Metaphysik in der zeitgenössischen Hegel Rezeption". In: *Hegel Jahrbuch* 2011. No. 2, 400–405.

Ficara, Elena (2013): "Dialectic and Dialetheism". In: *History and Philosophy of Logic* 34. No. 1, 35–52.

Ficara, Elena (2014a): "Hegel's Glutty Negation". In: *History and Philosophy of Logic* 36. No. 1, 29–38.

Ficara, Elena (2014b): "Hegel on the Mathematical Infinite". In: *Siegener Beiträge zur Geschichte und Philosophie der Mathematik* 4, 59–65.

Ficara, Elena (2014c): "Logik und Metaphysik". In: Koch, Anton Friedrich/Schick, Friederike/Vieweg, Klaus/Wirsing, Claudia (eds.): *Hegel – 200 Jahre Wissenschaft der Logik*. Hamburg: Meiner, 245–256.

Ficara, Elena (2015): "Contrariety and Contradiction. Hegel and the 'Berliner Aristotelismus'". In: *Hegel-Studien* 49, 39–55.

Ficara, Elena (2019): "Hegel and Priest on Revising Logic". In: *Graham Priest on Paraconsistency and Dialetheism*. Thomas Ferguson and Can Baskent (eds.). Cham: Springer, 59–72.

Ficara, Elena (2019): "Empowering Forms. Hegel's conception of 'form' and 'formal'". In: Bubbio, Diego/De Cesaris, Alessandro/Pagano, Maurizio/Weslati, Hager (eds.): *Hegel, Logic and Speculation*. London: Bloomsbury, 15–26.

Ficara, Elena (2020): "Truth and Incompatibility". In: Bouché, Gilles (ed.): *Reading Brandom. On A Spirit of Trust*. London: Routledge, 29–40.

Field, Hartry (2008): *Saving Truth from Paradox*, Oxford: Oxford University Press.

Findlay, John (1955): "Some Merits of Hegelianism". In: *Proceedings of the Aristotelian Society* 56, 1–24.

Findlay, John (1981), "Dialectic as Metabasis". In Becker, Werner/Essler, Wilhelm (eds.): *Konzepte der Dialektik*. Frankfurt a. M.: Klostermann, 132–139.

Frege, Gottlob (1884): *Grundlagen der Arithmetik. Eine logisch-mathematische Untersuchung über den Begriff der Zahl*. Breslau: Koebner.

Frege, Gottlob (1893): *Grundgesetze der Arithmetik* I. Jena: Verlag Hermann Pohle.

Frege, Gottlob (1956): "The Thought. A Logical Inquiry". In: *Mind* 65. No. 259, 289–311.

Frege, Gottlob (1979): "Logic (1897)". In: Hermes, Hans/Kambartel, Friedrich/Kaulbach, Friedrich (eds.): *Posthumous Writings*. Chicago: The University of Chicago Press.
Frege, Gottlob (2008): "Funktion und Begriff". In: Patzig, Günther (ed.): *Funktion, Begriff, Bedeutung. Fünf logische Studien*. Göttingen: Vandenhoeck & Ruprecht, 1–22.
Fulda, Hans Friedrich (1965): *Das Problem einer Einleitung in Hegels Wissenschaft der Logik*. Frankfurt a. M.: Klostermann.
Fulda, Hans Friedrich (1978): "Unzulängliche Bemerkungen zur Dialektik". In: Horstmann, Rolf-Peter (ed.): *Seminar: Dialektik in der Philosophie Hegels*. Frankfurt a. M.: Suhrkamp, 33–69.
Fulda, Hans Friedrich/Horstmann, Rolf-Peter (eds.) (1994): *Vernunftbegriffe in der Moderne*. Stuttgart: Klett-Cotta.
Fulda, Hans Friedrich (2006): "Methode und System bei Hegel. Das Logische, die Natur, der Geist als universale Bestimmung einer monistischen Philosophie". In: Fulda, Hans Friedrich/Krijnen Christian (eds.): *Systemphilosophie als Selbsterkenntnis. Hegel und der Neukantianismus*. Würzburg: Königshausen & Neumann, 25–50.
Gabbay, Dov/Woods, John (eds.) (2004): *Handbook of the History of Logic*. Vol. 3. Bonn: Elsevier.
Gabriel, Gottfried (2007): "Zur Einführung zum Kolloquium 'Der Weg nach Hegel: Trendelenburg, Lotze, Frege'". In: Hindrichs, Günnar/Bubner, Rüdiger (eds.): *Von der Logik zur Sprache. Stuttgarter Hegel-Kongress 2005*. Stuttgart: Klett-Cotta.
Gabriel, Gottfried (2008): "Wie formal ist die formale Logik? Friedrich Adolf Trendelenburg und Gottlob Frege". In: Bernhard, Peter/Peckhaus, Volker (eds.): *Methodisches Denken im Kontext. Festschrift für Christian Thiel*. Paderborn: Mentis, 115–131.
Gadamer, Hans-Georg (1976): *Hegel's Dialectics. Five Hermeneutical Studies*. New Haven and London: Yale University Press.
Gadamer, Hans-Georg (1987): *Hegel, Husserl, Heidegger*. Tübingen: Mohr.
Gerhard, Myriam (2015): *Hegel und die logische Frage*. Berlin New York: De Gruyter.
Glanzberg, Michael (2016): "Truth". In: Zalta, Ed (ed.): *The Stanford Encyclopedia of Philosophy* (Winter 2016 Edition). https://plato.stanford.edu/archives/win2016/entries/truth/, accessed 05/06/2020.
Glockner, Hermann (ed.) (1960): *Monadologie*. Stuttgart: Reclam.
Goble, Lou (ed.) (2007): *The Blackwell Guide to Philosophical Logic*. Oxford: Blackwell Publishing (first edition 2001).
Goclenius, Rudolph (1613): *Lexicon philosophicum*. Frankfurt: Becker.
Goethe, Johann Wolfgang (1808): *Faust. Eine Tragödie*. Tübingen: Cotta.
Günther, Gottard (1978): *Grundzüge einer neuen Theorie des Denkens in Hegels Logik*. Hamburg: Meiner (first edition 1933).
Haack, Susan (2006): *Philosophy of Logics*. Cambridge: Cambridge University Press (first edition 1978).
Haaparanta, Leila (ed.) (2009): *The Development of Modern Logic*. Oxford: Oxford University Press.
Hahn, Susan (2007): *Contradiction in Motion. Hegel's Organic Concept of Life and Value*, Ithaca: Cornell University Press.
Halbig, Christoph (2002): *Objektives Denken. Erkenntnistheorie und Philosophy of Mind in Hegels System*. Stuttgart-Bad Cannstatt: Frommann-Holzboog.

Halbig, Christoph/Quante, Michael/Siep, Ludwig (eds.) (2004): *Hegels Erbe*. Frankfurt a. M.: Suhrkamp.
Hamilton, Edith/Cairns, Huntington (eds.) (1961): *Plato: The Collected Dialogues*. Princeton: Princeton University Press.
Hammer, Espen (2007): *German Idealism. Contemporary Perspectives*. London: Routledge.
Hanna, Robert (1986): "From an ontological point of view. Hegel's critique of the common logic". In: *Review of Metaphysics* 40. No. 2, 305–338.
Havas, Katalin (1981): "Some Remarks on an Attempt at Formalizing Dialectical Logic". In: *Studies in Soviet Thought*, 22. No. 4, 257–264.
Hegel, Georg Wilhelm Friedrich (1892 ff.): *Lectures on the History of Philosophy*. Translated by Haldane, Elisabeth/Simson, Frances. London: Kegan Paul.
Hegel, Georg Wilhelm Friedrich (1969 ff.): *Werke in zwanzig Bänden. Theorie Werkausgabe. New edition on the basis of the Works of 1832–1845 edited by Eva Moldenhauer and Karl Markus Michel*. Frankfurt a. M.: Suhrkamp.
Hegel, Georg Wilhelm Friedrich (1969): *Hegel's Science of Logic*. Translation by Miller, Arthur. New York: Humanity Books.
Hegel, Georg Wilhelm Friedrich (1977): *Hegel's Phenomenology of Spirit*. Translation by Miller, Arthur (with analysis of the text and foreword by John Findlay). Oxford: Oxford University Press.
Hegel, Georg Wilhelm Friedrich (1991): *G. W. F. Hegel. The Encyclopaedia Logic. Part I of the Encyclopaedia of Philosophical Sciences with the Zusätze*. Translation by Geraets, Théodore/Suchting, Wal/Harris, Henry. Indianapolis: Hackett.
Hegel, Georg Wilhelm Friedrich (1992): *Vorlesungen über Logik und Metaphysik (Heidelberg 1817)*. Karen Gloy (ed.). Hamburg: Meiner.
Heidemann, Dietmar-Hermann (2007): *Der Begriff des Skeptizismus: Seine systematischen Formen, die pyrrhonische Skepsis und Hegels Herausforderung*. Berlin New York: De Gruyter.
Held, Klaus (1990): "Die Sophistik in Hegels Sicht". In: Riedel, Manfred (ed.): *Hegel und die antike Dialektik*. Frankfurt a. M.: Suhrkamp, 129–152.
Hempel, Carl Gustav (1977): "Zur Wahrheitstheorie des logischen Positivismus". In: Skirbekk Gunnar (ed.): *Wahrheitstheorien. Eine Auswahl aus den Diskussionen im 20. Jahrhundert*. Frankfurt a. M.: Suhrkamp, 96–108.
Henrich, Dieter (1965/66): "Hölderlin über Urteil und Sein. Eine Studie zur Entstehungsgeschichte des Idealismus". In: *Hölderlin-Jahrbuch*. 14, 73–96.
Henrich, Dieter (1978): "Formen der Negation in Hegels Logik". In: Horstmann, Rolf-Peter (ed.): *Seminar: Dialektik in der Philosophie Hegels*. Frankfurt a. M.: Suhrkamp, 213–229.
Hintikka, Jaakko (1981): "On Common Factors of Dialectics". In: Becker, Werner/Essler, Wilhelm (eds.): *Konzepte der Dialektik*. Frankfurt a. M.: Klostermann, 109–110.
Hintikka, Jaakko (2007): *Socratic Epistemology. Explorations of Knowledge-Seeking by Questioning*. Cambridge: Cambridge University Press.
Hodges, Wilfrid (2007): "Classical Logic I – First-Order Logic". In: Goble, Lou (ed.): *The Blackwell Guide to Philosophical Logic*. Oxford: Blackwell, 9–32.
Hölderlin, Johann Christian Friedrich (1961): *Der Tod des Empedokles. Aufsätze*. Friedrich Beißner (ed.). Stuttgart: Cotta.
Hösle, Vittorio (1998): *Hegels System. Der Idealismus der Subjektivität und das Problem der Intersubjektivität*. Hamburg: Meiner (first edition 1988).

Horn, Joachim Christian (1982): *Monade und Begriff. Der Weg von Leibniz zu Hegel.* Hamburg: Meiner.
Horn, Laurence (1989): *A Natural History of Negation.* Chicago: University of Chicago Press.
Horn, Laurence/Wansing, Heinrich (2016): "Negation". In: Zalta, Ed (ed.): *The Stanford Encyclopedia of Philosophy* (Spring 2016 Edition). http://plato.stanford.edu/archives/spr2016/entries/negation, accessed 05/06/2020.
Horsten, Leon/Pettigrew, Richard (eds.) (2014): *The Bloomsbury Companion to Philosophical Logic.* London New York: Bloomsbury.
Horstmann, Rolf-Peter (ed.) (1978): *Seminar: Dialektik in der Philosophie Hegels.* Frankfurt a. M.: Suhrkamp.
Horstmann, Rolf-Peter (1984): *Ontologie und Relationen: Hegel, Bradley, Russell und die Kontroverse über interne und externe Beziehungen.* Hain: Athäneum.
Houlgate, Stephen (2018): "Das Sein. Zweyter Abschnitt. Die Quantität". In: Quante, Michael/Mooren, Nadine (eds.): *Kommentar zu Hegels Wissenschaft der Logik.* Hamburg: Meiner, 145–218.
Hylton, Peter (1990): *Russell, Idealism, and the Emergence of Analytic Philosophy.* Oxford: Clarendon.
Illetterati, Luca (2010): "Contradictio regula falsi? Intorno alla teoria hegeliana della contraddizione". In: Puppo, Federico (ed.): *La contradizion che nol consente.* Milano: Angeli, 85–114.
Illetterati, Luca (2014): "Limit and Contradiction in Hegel". In: Ficara, Elena (ed.): *Contradictions. Logic, History, Actuality.* Berlin New York: De Gruyter, 127–152.
Ilting, Karl-Heinz (1973): *Die Rechtsphilosophie von 1820 und Hegels Vorlesungen über Rechtsphilosophie.* Stuttgart: Frommann-Holzboog.
Jacobi, Friedrich Heinrich (1998 ff.): *Werke. Gesamtausgabe.* Klaus Hammacher/Walter Jäschke (eds.). Stuttgart: Frommann-Holzboog.
Jäschke, Walter (2012): "Ein Plädoyer für einen historischen Metaphysikbegriff". In: Gerhard, Myriam/Sell, Annette/De Vos, Lou (eds.): *Metaphysik und Metaphysikkritik in der Klassischen Deutschen Philosophie.* Hegel-Studien Beiheft 57. Hamburg: Meiner, 11–22.
James, William (1999): "Pragmatism's Conception of Truth". In: Blackburn, Simon/Simmons, Keith (eds.): *Truth.* Oxford: Oxford University Press, 53–68.
Jaquette, Dale (ed.) (2002): *Philosophy of Logic. An Anthology.* Oxford: Backwell.
Jaquette, Dale (ed.) (2007): *Handbook of the Philosophy of Science: Philosophy of Logic.* Amsterdam: Elsevier.
Joachim, Harold (1999): "The Nature of Truth". In: Blackburn, Simon/Simmons, Keith (eds.): *Truth.* Oxford: Oxford University Press, 46–52.
Käufer, Stephan (2005): "Hegel to Frege: Concepts and Conceptual Content in Nineteenth-Century Logic". In: *History of Philosophy Quarterly* 22. No. 3, 259–280.
Kambartel, Friedrich (1976): *Theorie und Begründung.* Frankfurt a. M.: Suhrkamp.
Kant, Immanuel (1900 ff.): *Immanuel Kant, Kant's gesammelte Schriften, hrsg. von der Königlich Preußischen (später Deutschen) Akademie der Wissenschaften,* Berlin (quoted as AA, followed by the indication of volume and page. The *Critique of Pure Reason* is quoted, as usual, as A – first edition – and B – second edition, followed by the indication of the page).
Kant, Immanuel (1996): *Theorie-Werkausgabe Immanuel Kant. Werke in 12 Bänden hrsg. von W. Weischedel.* Frankfurt a. M.: Suhrkamp.

Kirn, Michael (1985): *Der Computer und das Menschenbild der Philosophie. Leibniz' Monadologie und Hegels philosophisches System auf dem Prüfstand*. Stuttgart: Urachhaus.

Kneale, William/Kneale, Martha (2008): *The Development of Logic*. Oxford: Clarendon Press (first edition 1962).

Koch, Anton Friedrich (2014): *Die Evolution des logischen Raumes. Aufsätze zu Hegels Nichtstandard-Metaphysik*. Tübingen: Mohr Siebeck.

Koch, Anton Friedrich (2018): "Das Sein. Erster Abschnitt. Die Qualität". In: Quante, Michael/ Mooren, Nadine (eds.): *Kommentar zu Hegels Wissenschaft der Logik*. Hamburg: Meiner, 43–144.

Koch, Anton Friedrich/Oberauer, Alexander/Utz, Konrad (eds.) (2003): *Der Begriff als die Wahrheit. Zum Anspruch der Hegelschen "Subjektiven Logik"*. Paderborn: Schöningh.

Kreines, James (2017): "From Objectivity to the Absolute Idea in Hegel's Logic". In: Moyar, Dean (ed.): *The Oxford Handbook of Hegel*. Oxford: Oxford University Press, 310–338.

Krohn, Wolfgang (1972): *Die formale Logik in Hegels "Wissenschaft der Logik". Untersuchungen zur Schlußlehre*. München: Hanser.

Kulenkampff, Arend (1970): *Antinomie und Dialektik. Zur Funktion des Widerspruchs in der Philosophie*, Stuttgart: Metzler.

Kulenkampff, Arend (1981): "Bemerkungen zur dialektischen Methode". In: Becker, Werner/ Essler, Wilhelm (eds.): *Konzepte der Dialektik*. Frankfurt a.M.: Klostermann, 140–144.

Labarrière, Pierre-Jean (1984), "L'esprit absolu n'est pas l'absolu de l'esprit. De l'ontologique au logique", in D. Henrich/R. P. Horstmann (eds.), *Hegels Logik der Philosophie. Religion und Philosophie in der Theorie des absoluten Geistes*, Stuttgart: Klett-Cotta, 35–41.

Landucci, Sergio (1978): *La contraddizione in Hegel*. Firenze: La Nuova Italia.

Lau, Chong-Fuk (2004): *Hegels Urteilskritik. Systematische Untersuchungen zum Grundproblem der spekulativen Logik*. München: Fink.

Lebanidze, Giorgi (2019): *Hegel's Transcendental Ontology*. New York London: Lexington Books.

Lejeune, Guillaume (ed.) (2013): *La question de la logique dans l'Idéalisme allemand*, Hildesheim: Olms.

Litt, Theodor (1961): *Hegel. Versuch einer kritischen Erneuerung*. Heidelberg: Quelle & Meyer.

Littmann, Greg/Simmons, Keith (2004): "A Critique of Dialetheism". In: Priest, Graham/Beall, Jc/Armour-Garb, Brad (eds.): *The Law of Non-Contradiction*. Oxford: Oxford University Press, 314–335.

Lowe, Jonathan (2013): *The Forms of Thought*. Cambridge: Cambridge University Press.

Lukács, Georg (1973): *Der junge Hegel*. Frankfurt a.M.: Suhrkamp.

Mac Farlane, John (2000): *What Does it Mean to Say that Logic is Formal?*. PhD Dissertation: University of Pittsburgh.

Marconi, Diego (ed.) (1979a): *La formalizzazione della dialettica*. Torino: Rosenberg & Sellier.

Marconi, Diego (1979b), "Introduction". In: Marconi, Diego (ed.): *La formalizzazione della dialettica*. Torino: Rosenberg & Sellier, 9–84.

Marcuse, Herbert (1941): *Reason and Revolution. Hegel and the Rise of Social Theory*. London: Oxford University Press.

McDowell, John (1994): *Mind and World*. Cambridge Mass.: Harvard University Press.

McTaggart, John (2000): *Studies in the Hegelian Dialectic*. Kitchener: Batoche Books.

Merker, Nicolao (1951): *Le origini della logica hegeliana. Hegel a Jena*. Milano: Feltrinelli.

Merker, Nicolao (1996): "Forma". In: Merker, Nicolao: *Hegel. Dizionario delle idee*. Roma: Editori Riuniti, 92.
Michelet, Karl Ludwig (1871): "Die Dialektik und der Satz des Widerspruchs". In: *Der Gedanke* 8. No. 1, 24–41. English translation by Louis Soldan, *The Journal of Speculative Philosophy* 5. No. 4, 319–337.
Mignucci, Mario (1995): "L'interpretazione hegeliana della logica di Aristotele". In: Movia, Giancarlo (ed.): *Hegel e Aristotele. Atti del Convergno di Cagliari, 11–15 aprile 1994*. Cagliari: Edizioni AV, 29–50.
Milkov, Nikolay (2020): *Early Analytic Philosophy and the German Philosophical Tradition*. London New York: Bloomsbury Academic.
Miolli, Giovanna (2016): *Il pensiero della cosa. Wahrheit hegeliana e Identity Theory of Truth*. Trento: Verifiche.
Mittelstraß, Jürgen (ed.) (1980 ff.): *Enzyklopädie Philosophie und Wissenschaftstheorie*,. Mannheim/Wien/Zürich: B.I. Wissenschaftsverlag.
Moiso, Francesco (2002): *Goethe, la natura e le sue forme*. Milano: Mimesis.
Moretto, Antonio (1984): *Hegel e la matematica dell'infinito*. Trento: Verifiche.
Moyar, Dean (ed.) (2017): *The Oxford Handbook of Hegel*. Oxford: Oxford University Press.
Ng, Karen (2017): "From Actuality to Concept in Hegel's *Logic*". In: Moyar, Dean (ed.): *The Oxford Handbook of Hegel*. Oxford: Oxford University Press, 269–290.
Nicolin, Günther (1971): *Hegel in Berichten seiner Zeitgenossen*. Berlin: Akademie Verlag.
Nuzzo, Angelica (1992): *Logica e sistema. Sull'idea hegeliana di filosofia*. Genova: Pantograf.
Nuzzo, Angelica (ed.) (1993): *La logica e la metafisica di Hegel. Guida alla critica*, Roma: La Nuova Italia.
Nuzzo, Angelica (1995): "Zur logischen Bestimmung des ontologischen Gottesbeweises. Bemerkungen zum Begriff der Existenz im Anschluß an Hegel". In: *Hegel-Studien* 30, 105–120.
Nuzzo, Angelica (1997): "La logica". In: Cesa, Claudio (ed.): *Hegel*. Roma Bari: Laterza, 39–82.
Nuzzo, Angelica (2003): "Existenz 'im Begriff' und Existenz 'außer dem Begriff'. Die Objektivität von Hegels Subjektiver Logik". In: Koch, Anton Friedrich/Oberauer, Alexander/Utz, Konrad (eds.): *Der Begriff als die Wahrheit. Zum Anspruch der Hegelschen "Subjektiven Logik"*. Paderborn: Schöningh, 171–187.
Nuzzo, Angelica (ed.) (2010a): *Hegel and the Analytic Tradition*. New York: Continuum.
Nuzzo, Angelica (2010b): "Vagueness and Meaning Variance in Hegel's Logic". In: Nuzzo, Angelica (ed.): *Hegel and the Analytic Tradition*. New York: Continuum, 61–82.
Nuzzo, Angelica (2011): "Truth and Refutation in Hegel's *Begriffslogik*". In: Ficara, Elena (ed.): *Die Begründung der Philosophie im Deutschen Idealismus*. Würzburg: Königshausen & Neumann, 91–105.
Nuzzo, Angelica (2014): "Dialektisch-spekulative Logik und Transzendentalphilosophie". In: Koch, Anton Friedrich/Schick, Friedrike/Vieweg, Klaus/Wirsing, Claudia (eds.): *Hegel – 200 Jahre Wissenschaft der Logik*. Hamburg: Meiner, 257–273.
Osnovy marksistskoj filosofii (1958): Moskow: Institute of Philosophy of the Academy of Sciences of the Soviet Union.
Peckhaus, Volker (1997): *Logik, Mathesis universalis und allgemeine Wissenschaft. Leibniz und die Wiederentdeckung der formalen Logik im 19. Jahrhundert*. Berlin: Akademie Verlag.

Peckhaus, Volker (1999): "19th Century Logic Between Philosophy and Mathematics". In: *The Bulletin of Symbolic Logic* 5. No. 4, 433–450.

Peckhaus, Volker (2004): "Calculus Ratiocinator versus Characteristica Universalis? The Two Traditions in Logic, Revisited". In: *History and Philosophy of Logic* 25, 3–14.

Peckhaus, Volker (2007). "Gegen 'neue unerlaubte Amalgamationen der Logik'. Die nachhegelsche Suche nach einem neuen Paradigma in der Logik". In: Hindrichs, Günnar/Bubner Rüdiger (eds.): *Von der Logik zur Sprache. Stuttgarter Hegel-Kongress 2005*. Stuttgart: Klett-Cotta, 241–255.

Peckhaus, Volker (2013), "Logik und Metaphysik bei Adolf Trendelenburg". In: Lejeune, Guillaume (ed.): *La question de la logique dans l'Idéalisme allemand*. Hildesheim: Olms, 283–296.

Peirce, Charles Sanders (1868): "Nominalism versus Realism". In: *Journal of Speculative Philosophy* 2, 57–61.

Perelda, Federico (2003): *Hegel e Russell. Ontologia tra moderno e contemporaneo*. Padova: Il Poligrafo.

Pinkard, Terry (1996): "What Is The Non-Metaphysical Reading Of Hegel? A Reply To F. C. Beiser". In: *Bullettin of the Hegel Society of Great Britain* 34, 13–20.

Pinkard, Terry (2002): *German Philosophy 1760–1860. The Legacy of Idealism*. Cambridge: Cambridge University Press.

Pinkard, Terry (2003): "Objektivität und Wahrheit innerhalb einer subjektiven Logik". In: Koch, Anton Friedrich/Oberauer, Alexander/Utz, Konrad (eds.): *Der Begriff als die Wahrheit. Zum Anspruch der Hegelschen "Subjektiven Logik"*. Paderborn: Schöningh, 119–134.

Pippin, Robert (1989): *Hegel's Idealism. The Satisfactions of Self-Consciousness*. Cambridge: Cambridge University Press.

Pippin, Robert (2014): "Die Logik der Negation bei Hegel". In: Koch, Anton Friedrich/Schick, Friederike/Vieweg, Klaus/Wirsing, Claudia (eds.): *Hegel – 200 Jahre Wissenschaft der Logik*. Hamburg: Meiner, 87–110.

Pippin, Robert (2016): *Die Aktualität des Deutschen Idealismus*. Berlin: Suhrkamp.

Pippin, Robert (2017): "Hegel on Logic as Metaphysics". In: Moyar, Dean (ed.): *The Oxford Handbook of Hegel*. Oxford: Oxford University Press, 199–218.

Pippin, Robert (2019): *Hegel's realm of shadows: logic as metaphysics in the science of logic*. Chicago: University of Chicago Press.

Pöggeler, Otto (1970): "Dialektik und Topik". In: Bubner, Rüdiger/Cramer, Konrad/Wiehl, Reiner (eds.): *Hermeneutik und Dialektik*. Tübingen: Mohr, 273–310.

Pöggeler, Otto (1981): "Schillers Antagonismus und Hegels Dialektik", in Becker, Werner/Essler Wilhelm (eds.): *Konzepte der Dialektik*. Frankfurt a. M.: Klostermann, 42–45.

Pöggeler, Otto (1990): "Die Ausbildung der spekulativen Dialektik in Hegels Begegnung mit der Antike". In: Riedel, Manfred (ed.): *Hegel und die antike Dialektik*. Frankfurt a. M.: Suhrkamp, 42–64.

Popper, Karl-Raimund (1965): *Conjectures and Refutations. The Growth of Scientific Knowledge*. London: Routledge.

Priest, Graham/Routley, Richard (1984): *On Paraconsistency*. Australian National University: Publications of the Philosophy Department, Research School of Social Sciences.

Priest, Graham (1989): "Dialectic and Dialetheic". In: *Science and Society* 53. No. 4, 388–415.

Priest, Graham (2002): *Beyond the Limits of Thought*. Oxford: Oxford University Press (first edition 1995).
Priest, Graham (2004): "What's so Bad about Contradictions?". In: Priest, Graham/Beall, Jc/Armour-Garb, Brad (eds.): *The Law of Non-Contradiction*. Oxford: Oxford University Press, 23–40.
Priest, Graham (2006): *Doubt Truth to Be a Liar*. Oxford: Oxford University Press.
Priest, Graham (2015): "19th Century German Logic". In: Forster Michael (ed.): *Oxford Handbook of German Philosophy in the Nineteenth Century*. Oxford: Oxford University Press, 398–415.
Puntel, Lorenz (1996): "Lässt sich der Begriff der Dialektik klären?". In: *Journal for General Philosophy of Science/Zeitschrift für allgemeine Wissenschaftstheorie* 27. No. 1, 131–165.
Puntel, Lorenz (2005): "Hegels Wahrheitskonzeption: Kritische Rekonstruktion und eine 'analytische' Alternative", In: *Internationales Jahrbuch des deutschen Idealismus* 3, 208–242.
Quante, Michael (2018): "Die Lehre vom Wesen. Erster Abschnitt. Das Wesen als Reflexion in ihm selbst". In: Quante, Michael/Mooren, Nadine (eds.): *Kommentar zu Hegels Wissenschaft der Logik*. Hamburg: Meiner, 275–324.
Quante, Michael/Mooren, Nadine (eds.) (2018): *Kommentar zu Hegels Wissenschaft der Logik*. Hamburg: Meiner, 145–218.
Quine, William Van Orman (1948): "On What There Is". In: *Revue Internationale de Philosophie*, later printed in: Willard Van Orman Quine: *From a Logical Point of View*, Cambridge Mass.: Cambridge University Press 2003, 1–19.
Quine, William Van Orman (1986): *Philosophy of Logic*, Cambridge Mass.: Harvard University Press (first edition 1970).
Rapp, Christoph/Wagner, Tim (2004): "Einleitung". In: Rapp, Christoph/Wagner, Tim (eds.): *Aristoteles' Topik*. Stuttgart: Reclam, 7–42.
Read, Stephen (1995): *Thinking About Logic*. Oxford: Oxford University Press.
Redding, Paul (2007): *Analytic Philosophy and the Return of Hegelian Thought*. Cambridge: Cambridge University Press.
Redding, Paul (2014): "The Role of Logic 'commonly so called' in Hegel's *Science of Logic*". In: *British Journal for the History of Philosophy* 22. No. 2, 281–301.
Redding, Paul (2017): "Subjective Logic and the Unity of Thought and Being: Hegel's Logical Reconstruction of Aristotle's Speculative Empiricism". In: *Logik/Logic. Internationales Jahrbuch des Deutschen Idealismus*. Dina Emundts/Sally Sedgwick (eds.). Berlin New York: De Gruyter, 165–188.
Restall, Greg (2006): *Logic. An Introduction*. London New York: Routledge.
Riedel, Manfred (ed.) (1990a): *Hegel und die antike Dialektik*, Frankfurt a. M.: Suhrkamp.
Riedel, Manfred (1990b): "Dialektik des Logos? Hegels Zugang zum 'ältesten Alten' der Philosophie". In: Riedel, Manfred (ed.): *Hegel und die antike Dialektik*. Frankfurt a. M.: Suhrkamp, 13–41.
Ritter, Joachim/Gründer, Karl/Gabriel, Gottfried (eds.) (1971ff.): *Historisches Wörterbuch der Philosophie*. Basel: Schwabe.
Rockmore, Tom (2005): *Hegel, Idealism, and Analytic Philosophy*. New Haven: Yale University Press.
Rockmore, Tom (2010): "Some Recent Analytic 'Realist' Readings of Hegel". In: Nuzzo, Angelica (ed.): *Hegel and the Analytic Tradition*. New York: Continuum, 158–172.

Rosen, Stanley (1990): "Hegel und der eleatische Fremde". In: Riedel, Manfred (ed.): *Hegel und die antike Dialektik*. Frankfurt a. M.: Suhrkamp, 153–168.

Routley, Richard/Meyer, Robert (1976): "Dialectical Logic, Classical Logic, and the Consistency of the World". In: *Studies in Soviet Thought* 16. No. 1/2, 1–25.

Routley, Richard/Meyer, Robert (1979): "Logica dialettica, logica classica e non-contraddittorietà del mondo". In: Marconi, Diego (ed.): *La formalizzazione della dialettica*. Torino: Rosenberg & Sellier, 324–353.

Routley, Richard (1978): "Semantics for Connexive Logics. I". In: *Studia Logica* 37, 393–412.

Routley, Richard/Priest, Graham (1984): *On Paraconsistency*, Logic Group/Department of Philosophy: University of Western Australia.

Ruggiu, Luigi/Testa, Italo (eds.) (2003): *Hegel Contemporaneo: la ricezione americana di Hegel a confronto con la tradizione europea*. Milano: Guerini.

Russell, Bertrand (1906): "The Theory of Implication". In: *American Journal of Mathematics* 28, 159–202.

Russell, Bertrand (1906–1907): "On the Nature of Truth". In: *Proceedings of the Aristotelian Society* 7. No. 1, 28–49.

Russell, Bertrand (1999): "William James's Conception of Truth". In: Blackburn, Simon/Simmons Keith (eds.): *Truth*. Oxford: Oxford University Press, 69–82.

Russell, Bertrand (2009): *Our Knowledge of the External World*. London: Routledge (first pulished in 1914).

Russell, Bertrand (2010): *The Philosophy of Logical Atomism*. London and New York: Routledge.

Sacchetto, Mario (1993): *La logica di Hegel e il problema della dialettica nel pensiero contemporaneo*. Torino: Paravia.

Sainsbury, Mark (2001): *Logical Forms. An Introduction to Philosophical Logic*. Oxford: Blackwell (first edition 1991).

Sainsbury, Mark (2009): *Paradoxes*. Cambridge: Cambridge University Press (first edition 1988).

Sandkaulen, Birgit (2019): *Jacobis Philosophie. Über den Widerspruch zwischen System und Freiheit*. Hamburg: Meiner.

Sarlemijn, Andries (1971): *Hegelsche Dialektik*. Berlin New York: De Gruyter.

Schäfer, Rainer (2001): *Die Dialektik und ihre besonderen Formen in Hegels Logik*, Hamburg: Meiner.

Schäfer, Rainer (2013): "Die syllogistische Genese des Widerspruchs in der absoluten Idee in Hegels Logik", In: *Teoria* 33. No. 1, 265–282.

Schick, Friederike (2003): "Begriff und Mangel des formellen Schliessens. Hegels Kritik des Verstandesschlusses". In: Koch, Anton Friedrich/Oberauer, Alexander/Utz, Konrad (eds.): *Der Begriff als die Wahrheit. Zum Anspruch der Hegelschen "Subjektiven Logik"*. Paderborn: Schöningh, 85–100.

Schick, Friederike (2018): "Die Lehre vom Begriff. Erster Abschnitt. Die Subjektivität". In: Quante, Michael/Mooren, Nadine (eds.): *Kommentar zu Hegels Wissenschaft der Logik*. Hamburg: Meiner, 457–558.

Schick, Stefan (2010): *Contradictio regula veri*. Hamburg: Meiner.

Schnädelbach, Herbert (1993): "Hegels Lehre von der Wahrheit". Berlin: Antrittsvorlesung vom 26. Mai 1993 an der Humboldt Universität zu Berlin.

Scholz, Heinrich (1959): *Abriß der Geschichte der Logik*. Freiburg München: Alber.

Sedgwick, Sally (2012): *Hegel's Critique of Kant. From Dichotomy to Identity*, Oxford: Oxford University Press.
Sextus Empiricus (1985): *Grundriß der pyrrhonischen Skepsis*. Frankfurt a. M.: Suhrkamp.
Shapiro, Stewart (2004): "Simple Truth, Contradiction, and Consistency". In: Priest, Graham/Beall, Jc/Armour-Garb, Brad (eds.): *The Law of Non-Contradiction*. Oxford: Oxford University Press, 336–354.
Sider, Ted (2010): *Logic for Philosophy*. Oxford: Oxford University Press.
Siep, Ludwig (2018): "Die Lehre vom Begriff. Dritter Abschnitt. Die Idee". In: Quante, Michael/Mooren, Nadine (eds.): *Kommentar zu Hegels Wissenschaft der Logik*. Hamburg: Meiner, 651–798.
Spaventa, Bertrando (1972): *Le prime categorie della logica di Hegel*. In: *Opere*. Giovanni Gentile (ed.). Vol. I. Firenze: Sansoni.
Sprigge, Timothy (2002): "Idealism". In Gale, Richard (ed.): *The Blackwell Guide to Metaphysics*. Oxford: Blackwell, 219–241.
Stekeler-Weithofer, Pirmin (1992): *Hegels Analytische Philosophie. Die Wissenschaft der Logik als kritische Theorie der Bedeutung*. Paderborn: Schöningh.
Stekeler-Weithofer, Pirmin (2005): *Philosophie des Selbsbewusstseins. Hegels System als Formanalyse von Wissen und Autonomie*. Frankfurt a. M.: Suhrkamp.
Stekeler-Weithofer, Pirmin (2016): "Hegel wieder heimisch machen". In: *Philosophische Rundschau* 63. No. 1, 3–16.
Stekeler-Weithofer, Pirmin (2018): "Das Sein. Dritter Abschnitt. Das Maass". In: Quante, Michael/Mooren, Nadine (eds.): *Kommentar zu Hegels Wissenschaft der Logik*. Hamburg: Meiner, 219–274.
Stern, Robert (1993): "Did Hegel Hold an Identity Theory of Truth?". In: *Mind* 102. No. 408, 645–647.
Stern, Robert (1996): "Hegel, Scepticism and Transcendental Arguments". In: Horstmann, Rolf-Peter/Fulda, Hans Friedrich (eds.): *Skeptizismus und spekulatives Denken in der Philosophie Hegels*. Stuttgart: Klett-Cotta, 206–225
Stewart, Jon (ed.) (1996): *The Hegel Myths and Legends*. Evanston: Northwestern University Press.
Strube, Claudius (1973): *Das Problem einer hermeneutischen Logik*. PhD Dissertation: Universität zu Köln.
Stuhlmann-Laeisz, Rainer (1976): *Kants Logik. Eine Interpretation auf der Grundlage von Vorlesungen, veröffentlichten Werken und Nachlass*. Berlin New York: De Gruyter.
Tarski, Alfred (1999): "The Semantic Conception of Truth and the Foundations of Semantics". In: Blackburn, Simon/Simmons, Keith (eds.): *Truth*. Oxford: Oxford University Press, 115–143.
Taylor, Charles (1983): *Hegel*. Frankfurt a. M.: Suhrkamp.
Theunissen, Michael (1978): "Begriff und Realität. Hegels Aufhebung des metaphysischen Wahrheitsbegriffs". In: Horstmann, Rolf-Peter (ed.): *Seminar: Dialektik in der Philosophie Hegels*. Frankfurt a. M.: Suhrkamp, 324–359.
Thiel, Christian (1965): *Sinn und Bedeutung in der Logik Gottlob Freges*. Meisenheim am Glan: Hain.
Tolley, Clinton (2019): "Hegel's Conception of Thinking in His Logic". In: Lapointe, Sandra (ed.): *Logic from Kant to Russell*. London: Routledge, 73–100.

Toth, Imre (1987): "Mathematische Philosophie und Hegelsche Dialektik". In: Petry, Michael (ed.): *Hegel und die Naturwissenschaften*. Stuttgart-Bad Cannstatt: Frommann-Holzboog, 89–182.
Trendelenburg Friedrich Adolf (1842): *Die logische Frage in Hegel's System. Zwei Streitschriften*. Leipzig: Brockhaus.
Tripodi, Paolo (2015): *Storia della filosofia analitica*. Milano: Carocci.
Tugendhat, Ernst (1970): *Vorlesungen zur Einführung in die sprachanalytische Philosophie*, Frankfurt a. M.: Suhrkamp.
Tugendhat, Ernst/Wolf, Ursula (1993): *Logisch-semantische Propädeutik*. Stuttgart: Reclam (first edition 1983).
van Eemeren, Frans/Grootendorst, Rob (2004): *A Systematic Theory of Argumentation. The pragma-dialectical approach*. Cambridge: Cambridge University Press.
Varnier, Giuseppe (1987): "Lo scetticismo nell'evoluzione della dialettica. Sul suo significato logico e gnoseologico nel primo pensiero jenese di Hegel". In: *Giornale Critico Della Filosofia Italiana* 7. No. 2, 282–312.
Varnier, Giuseppe (1990): *Ragione, negatività, autocoscienza. La genesi della dialettica hegeliana a Jena tra teoria della conoscenza e razionalità assoluta*. Napoli: Guida.
Varzi, Achille (2001): *Parole, oggetti, eventi e altri argomenti di metafisica*. Roma: Carocci.
Varzi, Achille (2011): "On Doing Philosophy without Metaphysics". In: *Philosophical Perspectives* 25, 407–423.
Varzi, Achille (2014): "Logic, Ontological Neutrality, and the Law of Non-Contradiction". In: Ficara, Elena (ed.): *Contradictions. Logic, History, Actuality*. Berlin New York: De Gruyter, 53–80.
Verra, Valerio (ed.) (1976): *La dialettica nel pensiero contemporaneo*. Bologna: Il Mulino.
Verra, Valerio (ed.) (1981): *Hegel interprete di Kant*. Napoli: Prismi.
Verra, Valerio (2007). *Su Hegel*. Claudio Cesa (ed.). Bologna: Il Mulino.
Viellard Baron, Jean-Louis (2013): "Le devenir logique: négativité et contradiction". In: *Teoria* XXXIII/1, 49–68.
Vieweg, Klaus (1999): *Philosophie des Remis. Der junge Hegel und das Gespenst des Skepticismus*. München: Fink.
Vieweg, Klaus/Welsch Wolfgang (eds.) (2003): *Das Interesse des Denkens: Hegel aus heutiger Sicht*. München: Fink.
Vieweg, Klaus (2007): *Skepsis und Freiheit. Hegel über den Skeptizismus zwischen Philosophie und Literatur*. München: Fink.
Vikko, Risto (2009): "The Logic Question During the First Half of the Nineteenth Century". In: Haaparanta, Leila (ed.): *The Development of Modern Logic*. Oxford: Oxford University Press, 203–221.
von Hartmann, Eduard (1868): *Über die dialektische Methode*. Berlin: Dunker.
Wagner, Steffen (2011): *La filosofia pratica di F. A. Trendelenburg*. Napoli: Luciano.
Wansing, Heinrich (2007): "Negation". In: Goble, Lou (ed.): *The Blackwell Guide to Philosophical Logic*. Oxford: Blackwell Publishing, 415–436.
Wasek, Norbert (1990): "Hegel und die antike Dialektik. Bibliographie". In: Riedel, Manfred (ed.): *Hegel und die antike Dialektik*. Frankfurt a. M.: Suhrkamp, 275–283.
Wason, Philip/Johnson-Laird, Peter (1972): *Psychology of Reasoning: Structure and Content*. Cambridge, Mass.: Harvard University Press.
Wetter, Gustav (1958): *Dialectical Materialism*. London: Routledge and Kegan Paul.

Wiehl, Reiner (1966): "Über den Sinn der sinnlichen Gewißheit in Hegels Phänomenologie des Geistes". In: Gadamer, Hans-Georg (ed.): *Hegel-Tage, Royaumont 1964: Beiträge zur Deutung der Phäomenologie des Geistes*. Hegel-Studien Beiheft 3. Bonn: Bouvier, 103–134.
Wolff, Christian (1730): *Philosophia prima sive ontologia*. Frankfurt Leipzig: Renger.
Wolff, Michael (1981): *Der Begriff des Widerspruchs. Eine Studie zur Dialektik Kants und Hegels*, Königstein: Hain.
Wolff, Michael (1984): "Der Begriff des Widerspruchs in der 'Kritik der reinen Vernunft'. Zum Verhältnis von formaler und transzendentaler Logik". In: Tuschling, Burkhard (ed.): *Probleme der "Kritik der reinen Vernunft"*. Berlin New York: De Gruyter, 178–202.
Wolff, Michael (1986): "Über Hegels Lehre vom Widerspruch". In: Henrich, Dieter (ed.): *Hegels Wissenschaft der Logik. Formation und Rekonstruktion*. Stuttgart: Klett-Cotta, 107–128.
Zambrana, Rocío (2017): "Subjectivity in Hegel's Logic". In: Moyar, Dean (ed.): *The Oxford Handbook of Hegel*. Oxford: Oxford University Press, 291–309.

Index of Names

Abel, Günter V, 12, 71
Adorno, Theodor Wiesengrund 15, 146
Alston, William 112
Ameriks, Karl 1
Apostel, Leo 37, 71, 122, 158, 171, 196 f.
Armstrong, David Malet 110
Asmus, Conrad 120
Asmuth, Christoph 101, 127
Aubenque, Pierre 134

Baldwin, Thomas 80, 82, 99 f.
Baptist, Gabriella 17, 141
Barnes, Johnathan 31, 86, 112, 114, 121, 128, 136, 152
Barone, Francesco 31
Baum, Manfred 122, 127, 134, 183
Baumann, Charlotte 20
Baumgarten, Alexander 20
Beall, Jc V, 161 f., 173
Becker, Werner 149
Beiser, Frederick 19, 120
Bellissima, Fabio 151–154
Bencivenga, Ermanno 1, 3, 150
Berti, Enrico 3, 35, 124, 134–137, 166, 170, 187
Bertinetto, Alessandro 31
Berto, Franz 1, 3, 36, 72, 81 f., 103, 111, 122, 150, 162, 172, 179 f., 195, 197 f.
Bochenski, Joseph Maria 4
Bodammer, Theodor 85
Boeder, Heribert 124
Böhm, Marcus V
Boole, George 54, 71
Bordignon, Michela 1, 36, 74, 149 f., 170 f., 173, 181
Boyer, Carl 4
Brandom, Robert V, 1 f., 80, 82 f., 150, 171
Breidbach, Olaf 71
Bremer, Manuel 198
Brinkmann, Klaus 17, 141
Bubner, Rüdiger 124
Buchner, Hartmut 137 f.
Burbidge, John 1, 5

Burgess, John 36
Buridan 120
Burkhard, Berndt 170
Butler, Clark 1, 3, 66, 122, 149, 156, 159

Cairns, Huntington 86, 152
Campogiani, Marco 85
Caron, Maxence 12
Cassirer, Ernst 57
Chalmers, David 38
Cislaghi, Federica 71
Colombo, Enrico 170
Cook, Roy 34 f., 46, 154
Cortella, Lucio 180
Croce, Benedetto 35, 101, 146, 159

d'Agostini, Franca V, 1, 3, 7, 12, 21, 23, 33, 36, 68, 101, 103, 108, 111, 122, 131 f., 150, 151, 153 f., 173, 195
Dangel, Tobias 134
De Boer, Karin 19, 170
de Cesaris, Alessandro V
de Laurentiis, Allegra 19, 49
De Vos, Lou 44
Demmerling, Christoph 17
D'Hondt, Jacques 109
di Giovanni, George 12
Dilthey, Wilhelm 3
Dottori, Riccardo 12
Dummett, Michael 67
Düsing, Klaus V, 3, 9, 17, 20, 24, 32, 84 f., 91 f., 95, 122, 127, 129, 134, 141, 148, 174, 180, 183, 187
Dutilh Novaes, Catarina 46, 70, 121

Eckermann, Johann Peter 123, 136, 171
Emundts, Dina 1, 101 f.
Engelhard, Kristina 17, 141
Engelhardt, Tristram 19
Essler, Wilhelm 149
Euler 54, 57, 70

Ferrarin, Alfredo 23 f., 49, 134

Index of Names

Ferrater Mora, José 21
Ficara, Elena 5, 14, 19–21, 24, 28, 31, 36f., 44, 83, 170f., 181
Field, Hartry 195
Findlay, John 103f., 151, 158f.
Frege, Gottlob 1f., 6f., 12, 23, 28, 38, 52, 57, 67f., 71, 80, 84, 111, 120, 152, 170
Fulda, Hans Friedrich 12, 16, 148

Gabbay, Dov 5
Gabriel, Gottfried 2, 4f., 15, 24, 30, 45f., 57f., 66, 71, 84, 91f., 120, 179
Gadamer, Hans-Georg 1, 3, 9, 12, 17, 20, 28, 122, 124, 134, 148, 159
Gerhard, Myriam 24, 46, 71
Glanzberg, Michael 99
Glockner, Hermann 54
Goble, Lou 35f.
Goclenius, Rudolph 20
Goethe, Johann Wolfgang 57, 71, 123, 136
Grootendorst, Rob 124
Gründer, Karl 4, 15, 30, 45, 58, 66, 84, 91f., 120, 179
Günther, Gottard 35f., 148, 171

Haack, Susan 161
Haaparanta, Leila 5
Hagengruber, Ruth V
Hahn, Susan 170
Halbig, Christoph 23, 83
Hamilton, Edith 86, 152
Hammer, Espen 1
Hanna, Robert 32, 46f., 171
Havas, Katalin 173
Hegel, Georg Wilhelm Friedrich 1–28, 30–77, 79–117, 119–151, 153–197, 199–201
Heidemann, Dietmar-Hermann 9, 137
Held, Klaus 124
Heller, Ute V
Hempel, Carl Gustav 108f.
Henrich, Dieter 3, 92, 179f.
Hintikka, Jaakko 124, 148–150
Hodges, Wilfrid 31
Hölderlin, Johann Christian Friedrich 91f.
Horn, Joachim Christian 54, 105, 179
Horn, Laurence 54, 105, 179

Horsten, Leon 37
Horstmann, Rolf-Peter 16, 150
Hösle, Vittorio 81
Houlgate, Stephen 19
Hylton, Peter 34

Illetterati, Luca 170
Ilting, Karl-Heinz 109

Jacobi, Friedrich Heinrich 184
James, William 109
Jaquette, Dale 33–36
Jäschke, Walter 19
Joachim, Harold 80f., 108
Johnson-Laird, Peter 14

Kambartel, Friedrich 33
Kant, Immanuel 4, 8, 11, 13, 15–21, 23f., 26, 30–32, 44, 47, 49–51, 54, 57–63, 67, 75f., 83–85, 91, 97, 100, 102, 104f., 122f., 141–143, 147, 165, 181, 187
Käufer, Stephan 2, 52, 72
Kirn, Michael 54
Kneale, Martha 4, 30, 32, 123, 155f.
Kneale, William 4, 30, 32, 123, 155f.
Koch, Anton Friedrich 19, 32, 82, 96
Kreines, James 19
Krohn, Wolfgang 14, 32, 45f., 58, 66
Kulenkampff, Arend 148, 151, 158

Labarrière, Pierre-Jean 12
Lambert, Johann Heinrich 54, 57, 70
Landucci, Sergio 180
Lau, Chong-Fuk 81f., 85f., 92
Lebanidze, Giorgi 19
Lebock, Sarah V
Leibniz, Gottfried-Wilhelm 2, 5f., 47, 54f., 57, 70f., 75
Lejeune, Guillaume 31, 170
Litt, Theodor 35, 59, 66f., 71
Littmann, Greg 195
Lowe, Jonathan 7, 34
Lukács, Georg 3

Mac Farlane, John 46, 70
Manley, David 38

Marconi, Diego   3, 36, 71, 74, 122, 148–151, 156, 158, 171, 180
Marcuse, Herbert   3, 109
McDowell, John   150
McTaggart, John   170 f.
Merker, Nicolao   28, 170
Meyer, Robert   158, 171, 173
Michelet, Karl Ludwig   122 f., 129, 134, 148
Mignucci, Mario   24, 48, 121, 134, 165
Milkov, Nikolay   V, 34
Miolli, Giovanna   82
Mittelstraß, Jürgen   46, 66
Moiso, Francesco   71
Mooren, Nadine   1
Moretto, Antonio   37, 56
Moyar, Dean   1

Ng, Karen   101
Nicolin, Günther   123
Nuzzo, Angelica   V, 1, 12, 14, 31 f., 44, 46, 67, 74, 82, 87, 89, 101, 150

Oberauer, Alexander   96

Pagli, Paolo   151–154
Peckhaus, Volker   2, 4 f., 24, 37, 45 f., 54, 170
Peirce, Charles Sanders   170 f.
Perelda, Federico   180
Perler, Dominik   V
Pettigrew, Richard   37
Pinkard, Terry   19, 82 f., 96
Pippin, Robert   1, 19 f., 82, 171, 180
Pöggeler, Otto   3, 122, 134, 174, 187
Popper, Karl-Raimund   170 f.
Priest, Graham   V, 3, 5, 36, 122–124, 141, 148–150, 154–156, 161, 167, 171, 173, 182, 195
Puntel, Lorenz   80, 82, 107, 179

Quante, Michael   V, 1, 23, 83
Quine, William Van Orman   31, 38

Rapp, Christoph   134, 136
Read, Stephen   46, 70, 72, 161 f.
Redding, Paul   1–3, 24, 32, 46, 53, 82, 105, 120, 122, 150, 180

Restall, Greg   59, 120, 161 f.
Riedel, Manfred   3, 9, 20, 122
Ritter, Joachim   4, 15, 30, 45, 58, 66, 84, 91 f., 120, 179
Rockmore, Tom   1, 101, 103–105
Rosen, Stanley   127
Routley, Richard   5, 123 f., 149, 154–156, 158, 167, 171, 173, 182, 196
Ruggiu, Luigi   150
Russell, Bertrand   2, 6 f., 11, 28, 34 f., 37–40, 42, 44, 70, 95, 108–111, 152

Sacchetto, Mario   71
Sainsbury, Mark   7, 34 f., 70, 73, 133, 154
Sandkaulen, Birgit   184
Sarlemijn, Andries   156
Schäfer, Rainer   V, 3, 84, 159, 187
Schick, Friederike   46 f., 170
Schick, Stefan   46 f., 170
Schirmer, Christoph   V
Schnädelbach, Herbert   100
Scholz, Heinrich   58
Sedgwick, Sally   1, 17, 32, 141
Sextus Empiricus   9
Shapiro, Stewart   37, 195
Sider, Ted   162
Siep, Ludwig   5, 83
Simmons, Keith   195
Spaventa, Bertrando   170
Spinoza, Baruch   184–186, 189
Sprigge, Timothy   81
Stekeler-Weithofer, Pirmin   V, 1, 5, 16, 19, 28, 81, 102, 105, 150, 187
Stern, Robert   19, 80, 99 f.
Stewart, Jon   171
Stolzenberg, Jürgen   1
Strube, Claudius   31
Stuhlmann-Laeisz, Rainer   31

Tarski, Alfred   73, 112, 117, 120
Taylor, Charles   31
Testa, Italo   150
Theunissen, Michael   100
Thiel, Christian   5
Tolley, Clinton   5
Toth, Imre   171

Trendelenburg, Friedrich Adolf   2, 5, 24, 57, 71, 146, 170
Tripodi, Paolo   111
Tugendhat, Ernst   7, 38f., 58, 80, 85f.

Utz, Konrad   96

van Eemeren, Frans   124
Varnier, Giuseppe   137f.
Varzi, Achille   2, 21, 111
Verra, Valerio   3, 17, 20, 24, 49, 129, 138, 141, 170
Vieweg, Klaus   V, 3, 9, 15, 137f.
Vikko, Risto   5
von Hartmann, Eduard   129

Wagner, Steffen   134, 136, 170
Wagner, Tim   134, 136, 170

Wansing, Heinrich   179
Wasek, Norbert   122
Wason, Philip   14
Wasserman, Ryan   38
Weber, Mara   V
Wetter, Gustav   173
Weyand, Juliette   V
Weyand, Knut   V
Wiehl, Reiner   17
Wolf, Ursula   85f.
Wolff, Christian   20, 31, 38, 84, 91, 141, 170, 181
Wolff, Michael   20, 31, 38, 84, 91, 141, 170, 181
Woods, John   5

Zambrana, Rocío   19, 71

Index of Subjects

Antinomies   17, 141f., 151, 199

Berliner Aristotelismus   24

Common sense   15, 116
Concepts   1–3, 5, 12, 15, 19–21, 23, 28, 35, 38f., 49f., 52, 56, 60, 70, 72, 74, 80, 83, 85, 88, 90, 93f., 96, 100–103, 105, 111, 116, 120, 124–127, 130–134, 136, 142, 144, 146, 150f., 157f., 160, 165, 167, 179f., 183, 185, 189f., 194, 199
– and antinomies   17, 139, 141f, 154, 170, 173–178, 187, 199
– and categories   20f., 41, 49
– and definitions   65, 74, 96, 100, 114, 177
– and dialectic   56f., 67, 70f., 74, 77, 84, 101, 124–168, 173–178, 180–187, 194–198
– and ideas   26, 102f., 132
– and inferences   2, 13, 41f., 49, 51–53, 65, 74, 83, 149, 164–168, 183, 197, 201
– and judgements   38, 52, 80f., 92
– as essences of things   40, 46, 50f., 65, 75, 87
– as forms   17–29, 33, 38–42, 46, 49–52
– completeness of   28, 142, 185, 188
– containing different and opposite determinations   93, 146, 147
– contradictory   81, 131, 134f., 147, 157, 173–178, 185f., 196, 199–201
– dynamic nature of   28, 64, 66, 71f., 74f., 77, 80, 90, 181
– fluid nature of   55, 69f., 88, 92, 108, 142, 146, 157, 178, 199
– heterogeneous   92, 142, 146, 157, 178, 199
– internally differentiated   102, 131, 134f., 185
– life of   56
– movement of   5, 57, 84, 88, 90, 93, 121, 124f., 130, 148, 150, 159, 165

– philosophical   1f., 4f., 7, 9, 12, 14, 16f., 20, 26–28, 31, 34–36, 38, 42, 50, 56, 65, 71f., 76f., 80, 91, 94–96, 98–101, 104, 106, 112, 114, 123, 128, 130, 132, 134, 136, 148f., 151f., 156, 160, 163, 170, 184, 189f., 192, 194, 200
– pure   7, 11f., 21, 32, 34, 37, 44, 57f., 61, 67, 76, 85–88, 98, 102–104, 106, 110, 121, 124–128, 130, 133, 142, 165, 182, 185
– rational   16, 18f., 33, 35, 45, 50, 56, 81, 97f., 101–104, 106, 109, 111, 113, 117, 139f., 144, 146f., 157, 159, 163, 184, 189f., 192, 196, 200
– realisation of   83, 103
– reflexive   9, 64, 76, 89, 93, 102, 121, 165, 186
– self-referential   87, 140, 165
– semantic   28, 36f., 41, 71, 74, 80, 102, 105, 107, 115f., 121, 124, 132f., 148, 151, 158f., 161–163, 167, 173, 180, 195, 199
– the concept [*der Begriff*]   28, 82, 127
– versus representations [*Vorstellungen*]   96
Consequentia mirabilis   5, 151, 153, 155, 166f.
– the marvellous fact (*das Wunderbare*) in Hegel   130, 153
Contradiction   6, 10, 17f., 74, 81, 125f., 128–130, 132, 135, 137f., 141–143, 145f., 148f., 159, 161, 165f., 170–175, 183f., 194–201
– and completeness   18, 132, 162
– principle of completion   174
– and contrariety   137, 146, 183
– and the Opposition Principle (O)   190, 192
– and the Principle of Contradiction (C)   188
– as the form of truth   7–9, 51, 63, 187f., 194
– as the norm of truth   8, 40, 159, 200

Index of Subjects — 223

- biconditional contradictions   149, 173, 197
- and simplification   194, 197
- explosion   171, 195–198, 200f.
  - dialectic and dialetheism   201, 150, 194f.
- true contradictions   124, 128, 150, 154, 167, 173, 175, 181, 194f., 197f., 200f.

Dialectic   10, 15, 24, 33, 35, 56, 70f., 81, 101, 113, 121f., 125–129, 132, 135, 137, 145–150, 156, 158f., 170f., 179–181, 196
- and *reductio ad absurdum*   149, 153, 166
- and scepticism   145–147
- and sophistry   124, 127f., 132, 145–147, 156, 165, 175
- and the "three sides of *das Logische*"   144–147
- as the art of the dialogue   165
- as the logic of contradiction   121, 124, 165f.
- as the movement of pure concepts   88, 121, 125, 130, 165
- dialectical inferences   3, 5, 10, 121, 133–136, 144, 148–151, 153–155, 158–161, 165, 167, 171, 178, 183, 201
  - and *consequentia mirabilis* arguments   151–157
  - and paradoxical deductions   151, 154f., 160, 167
- external dialectic versus immanent dialectic   125, 145, 147, 156, 165
- formalisation of   10, 34, 70–82
- Hegel's dialectic and Aristotle's *Topics*   134–137
- immanent dialectic with negative result versus immanent dialectic with positive result   126, 156, 165f.
Doppelsatz   102, 104, 111f., 117

Elenchos   128f., 156f.
Empiricism   45, 48f., 59f., 104f.
- speculative empiricism of Aristotle   45, 49, 105

Form of truth   7–9, 51, 63, 88f., 144, 187f., 194
Formal   1f., 6, 8, 10, 18, 21, 23, 31f., 35f., 38f., 42, 44–46, 50f., 53f., 57–59, 61f., 64–72, 74, 76f., 94, 101f., 137f., 140f., 144, 154, 157f., 160–164, 166, 171, 188–190, 196f., 200
- as general   57–59
- as pertaining to forms   46
- as pertaining to rules   46
- as pertaining to the structure of sentences or arguments   46
- as reflexive   58, 76
- as symbolic   46, 54–57
- as the result of abstraction   46, 58
- as the result of semantic ascent   58, 76
- as universal and necessary   60–62
Formalistic argument (FA)   44, 61–63, 66, 76f.
Forms see Logical forms

Idealism   19–21, 31, 44f., 81, 101, 123
- and realism   101
- metaphysical   44f.
- transcendental   20f.
Ideas   26, 102–105, 132
- and rationality   102
- reality of   103f.
Infinite   37, 50, 56f., 59, 75, 81, 126f., 142f., 179, 185
- bad infinite   56
- genuine infinite   56
- mathematical infinite   56f.

Judgement (*Urteil*)   1, 25, 63, 80, 84–88, 91f., 94f., 99, 114–116

Law of Excluded Middle   132, 137, 140, 187–193
Law of Non-Contradiction   187–193
Liar Paradox   133f., 154, 167, 175
Logic   1–28, 30–38, 40–42, 44–55, 57–72, 75–77, 80, 83–85, 87, 89, 91–93, 97f., 102f., 107–109, 111, 113–115, 119–124, 126, 129, 133–135, 137, 144, 148, 150, 152, 155, 158, 161–163,

165f., 170f., 173, 176, 180, 182f., 187, 190f., 193, 199f.
- alethic logic   1
- and education   45
- and truth 1, 7, 10, 30–46, 61
- classical logic   36, 71f., 151, 163, 172, 199
- common logic [gewöhnliche Logik]   31
- conceptual logic   27–29, 179
- das Logische   9, 12–14, 17f., 20f., 23f., 28–30, 40f., 49, 64, 71, 144, 147
- dialectical logic   1f., 4, 9, 15, 26, 32, 42, 52, 66f., 77, 84, 121, 138, 150, 158f., 163, 172, 196, 201
- die logische Frage   5, 24, 170
- formal logic   4–6, 15, 21, 30–32, 36, 38, 44, 46–53, 57, 61, 65–67, 70f., 74–77, 138, 149
  - and transcendental logic   31–33, 57, 60
  - as calculus ratiocinator   54–57
  - as lingua characteristica   54–57
  - critique of   44–47, 60–63, 66–69
- Hegel's logic   1–3, 5, 7f., 10, 15f., 19, 24, 26, 28, 30, 32f., 35, 37f., 41f., 44f., 47, 54, 66, 69–72, 77, 85, 101, 115, 120, 131, 146, 148, 150, 160, 163, 165, 170, 174, 179–181, 194, 196
  - and formalisation   70–74
  - and the history of logic 4–7
  - as conceptual analysis   38–40
  - as non-classical logic   36f., 71f., 151, 163, 172, 180, 197
  - as philosophical logic   7f., 30–40
  - as rebuilding traditional logic   46
- history of logic   3–5, 10, 30, 41, 46, 122, 141, 149–151, 159, 166, 179
- intellectual logic (Verstandeslogik)   15f., 18, 32, 36, 41, 45, 75
- logica naturalis and logica scholastica   13
- logic as science   44
- logic as theory   2, 12–15, 23f., 28f., 31–33, 35, 41, 49, 66
- mathematical logic   2, 4, 33, 37
- natural logic   5, 7f., 12–16, 24–26, 32, 34, 37, 41f., 44, 48, 74, 88, 106

- non-classical logic   36f., 71f., 151, 163, 172, 180, 197
  - paraconsistent logic   5, 174, 194, 196
  - relevant logic   72, 163, 198
- philosophical logic   1–4, 6f., 10f., 26–28, 30, 33–42, 44, 46, 51, 69f., 72, 91, 115, 117, 120, 122, 137, 141, 162, 172, 179
- logic of philosophy   35, 135, 166
- rational logic (Vernunftlogik)   9, 15f., 18, 30, 41, 138, 200
- speculative logic   9, 31, 38, 41, 80, 181
- syllogistic logic   45
- transcendental logic   5, 20, 31f., 47, 57, 60, 76
- and ontology   20f., 23, 41, 49
- universal and necessary character of   60
Logical forms   7, 10–14, 17–29, 33f., 36, 38–42, 44, 46f., 49–53, 56, 60, 63–68, 70, 72f., 75, 77
- and content   44f., 62, 68, 71, 76, 84, 95, 167
- as deposited in natural language and reasoning   7, 41, 76, 124, 148f.
- as modalities of thought   76, 102
- as norms of thought   28f., 76
- as revealing the essence of things   77
- as seeds in plants   159
- as truth-implying   67f.
- forms of reality   29, 42, 124
- forms of thought   7f., 11–13, 17, 21, 23, 26, 28, 31, 33, 40–42, 44, 48–50, 53, 58, 62, 64, 66, 75, 87, 124, 135, 158, 166
- forms of truth   1f., 8, 39, 45, 51f., 54, 62f., 75, 77, 163
- generative   71
- organic   2, 57, 68, 71f., 89
- power of   127
- self-revising   28, 66
- syllogistic forms   6, 46, 49, 51

Metaphysics   1f., 4f., 10, 12f., 15, 19–26, 30, 32f., 41f., 49, 64, 66, 68, 83, 88, 97, 104, 106, 111f., 114, 120, 136–138, 143, 150, 152
- das Metaphysische   20f., 23, 29

– metaphysics as theory   23 f., 29
– natural metaphysics   13, 24 f., 42

Natural reasoning   14, 70
– errors of   14
Negation   4, 6, 76, 80, 89 f., 113, 125, 130 f., 141, 145 f., 148, 151–157, 171–174, 177–187, 190, 192, 194, 197, 199–201
– and concepts   17, 52, 83, 111, 136
– as contradictory forming operator   180 f., 199
– and scepticism 182
– as cancellation   182
– connexivist   182
– determinate negation   74, 88, 137, 139, 180–182, 184 f., 191
– double negation   10, 113, 171 f., 179 f., 183 f., 200
– internal   40, 51, 55, 92, 126, 139, 165 f., 180 f., 194, 199 f.
– iteration of negation in dialectic 194, 200
– Law of Dialectical Double Negation (DDN) 185
– negativity   4, 139, 179, 181
– predicative negation   179 f.
– propositional negation   179 f.

Objective thought   20, 22 f.
Ontology   19–21, 23, 38, 41, 49, 111

Principle of Identity (I)   15, 187 f., 190

Reality (*Wirklichkeit*)   8, 18 f., 21–25, 28 f., 33, 40–42, 44–46, 58, 62, 68, 76, 81 f., 86, 94, 97–99, 101–112, 115–117, 144, 162 f., 166, 176, 181, 188
– and rationality 81, 101–106, 144, 171
– sentence-like nature of reality   110 f.
Reflexive thought   9, 16, 75 f., 86, 121

Scepticism   9, 18, 113, 137–139, 141, 145, 153, 157, 182, 187, 189 f., 200
– and logic 113, 137–139
– and philosophy   137–139
– sceptical principle (S)   9, 138, 182, 189 f.

Semantic ascent   46, 58, 76, 130, 138, 151, 153
Sentence (*Satz*)   2, 15, 38, 73 f., 80 f., 84–88, 90 f., 93–97, 100–105, 108–112, 114–117, 131, 133, 137 f., 141, 151, 153–155, 173 f., 176–180, 182, 184, 189–191, 196, 198, 200
– speculative sentence (*spekulativer Satz*) 84 f., 91, 95, 187
Syntax   28, 158
– interplay with semantics   28, 158

Tarski's Schema   112, 117, 154 f.
– and Aristotle's definition of truth   117
– and Hegel's *Doppelsatz*   102, 104, 111 f., 117
Transcendental philosophy   16 f., 19, 21, 57
True   72 f., 80, 87–91, 102, 112, 114, 125, 133, 136 f.,
– as the process   87–91
– as the whole   87–91
– meaning of "true" 97–106
Truth   1–4, 6–11, 14, 18, 23, 25, 27 f., 31 f., 36–45, 49–53, 56, 59, 61–63, 65–69, 72–74, 76 f., 79–91, 93–100, 102 f., 107–117, 120–124, 126 f., 129, 131–134, 136–139, 144, 148 f., 151–168, 171–176, 178, 181–184, 186–188, 190 f., 194 f., 197, 200
– and certainty   97 f.,
– and *episteme*   88, 114, 116
– and validity   74, 160–164
– as coherence   80–82, 99, 107–109
– as correspondence   45, 62, 81, 97, 109 f., 114–116, 162
– as property of sentences   80, 99 f., 116
– as property of things   80, 89
– conceptual truth   83 f., 89, 93 f., 96, 111, 163, 168, 200
– correctness (*Richtigkeit*) versus truth (*Wahrheit*)   65, 91, 93 f., 99, 116, 194
– philosophical truth   87, 94, 132, 200
– pragmatistic conception of truth   81, 83, 107–109, 116 f.
– speculative truth   55, 89, 115 f., 177 f.
– transcendental truth   39
Truth-bearers   84–96

Validity 1, 6, 9f., 14f., 18, 37, 41, 72f., 119–122, 134, 140, 144f., 147f., 150f., 158f., 161–163, 165–167, 187
– and dialectical logic 147–168
– in Hegel and Aristotle 48–52, 121f., 134–137, 165
– proof-theoretic and model-theoretic validity 161, 167
– semantic validity 121, 158f., 161–163, 167
Vereinigung 4, 128, 167, 169f., 172–176, 178–180, 186f., 189, 194–197, 199–201

www.ingramcontent.com/pod-product-compliance
Lightning Source LLC
Chambersburg PA
CBHW032058230426
43662CB00035B/596